Effect of plant characteristics on host plant selection and larval performance of specialist insect pests on Brassicaceae

Effect of plant characteristics on host plant selection and larval performance of specialist insect pests on Brassicaceae

Dissertation
zur Erlangung des Doktorgrades
der Fakultät für Agrarwissenschaften
der Georg-August-Universität Göttingen

vorgelegt von

Alexander Döring
geboren in Fritzlar

Göttingen, Februar 2012

Bibliografische Information der Deutschen Nationalbibliothek
Die Deutsche Nationalbibliothek verzeichnet diese Publikation in der Deutschen Nationalbibliografie; detaillierte bibliografische Daten sind im Internet über http://dnb.d-nb.de abrufbar.
1. Aufl. - Göttingen : Cuvillier, 2012
Zugl.: Göttingen, Univ., Diss., 2012

978-3-95404-262-3

D7

1. Referent: Prof. Dr. Stefan Vidal

2. Korreferent: Prof. Dr. Heiko Becker

Tag der mündlichen Prüfung: 09.02.2012

Nonnenstieg 8, 37075 Göttingen
Telefon: 0551-54724-0
Telefax: 0551-54724-21
www.cuvillier.de

1. Auflage, 2012
Gedruckt auf säurefreiem Papier

978-3-95404-262-3

Table of contents

Chapter I: General introduction

Oilseed rape

Oilseed rape (*Brassica napus* L.) is the most important oilseed crop in Germany, in 2009 covering an acreage of 1.47 million hectares (Eurostat, 2009). Due to its higher yield potential winter oilseed rape is grown much more commonly than to spring rape (Kimber and McGregor, 1995). In 2010 winter oilseed rape yielded in average 39 dt/ha in Germany (Anonymus, 2011). The oil extracted from the harvested seeds is commonly used for human nutrition or for the production of biodiesel (Alford, 2003). The seed meal is used in animal nutrition because of its high content of protein and essential amino acids (Bell, 1995). Today most often cultivars with low content of erucic acid and glucosinolates in the seeds are grown, also known as 'canola quality' or 'double low quality' (Bell, 1995; Alford, 2003). Due to the low content of erucic acid, the oil is suitable for human nutrition. The low level of glucosinolates in the meal increases its suitability for animal nutrition (Bell, 1995; Uppström, 1995).

Oilseed rape pests in Germany

In Germany oilseed rape is attacked by a wide range of insect pests during the whole growing season (Alford et al., 2003). Economic damage is most often caused by only six insect species: cabbage stem flea beetle (*Psylliodes chrysocephala*), rape stem weevil (*Ceutorhynchus napi*), cabbage stem weevil (*Ceutorhynchus pallidactylus*), pollen beetle (*Meligethes aeneus*), cabbage seed weevil (*Ceutorhynchus obstrictus*) and brassica pod midge (*Dasineura brassicae*) (Williams, 2010).

The cabbage stem flea beetle is a common pest particularly in maritime regions of northern Europe (Williams, 2010). The adult beetles invade the oilseed rape fields during September after emergence of the seedlings (Schulz, 1992). As the ovaries of the females are immature at this stage, they conduct a maturity feeding over a period of two weeks (Williams, 2010). Gravid females lay their eggs into the upper soil layer near the roots of host plants (Kaufmann, 1941; Saringer, 1984). Neonate larvae bore into petioles and stems of oilseed rape plants where they feed until completion of larval development (Kaufmann, 1941; Godan, 1951). The first larvae start to pupate in soil already in winter but the majority leave their hosts plant in spring (Williams, 2010). New generation beetles emerge from soil in early summer and feed on oilseed rape until they migrate to hibernation sites for summer aestivation (Kaufmann, 1941; Williams & Carden, 1961). In addition to oilseed rape, many other brassicaceous plant species like runch (*Raphanus raphanistrum*), cabbage (*Brassica oleracea*) or turnip rape (*Brassica rapa*) are known to be host plants for cabbage stem flea beetles (Godan, 1951). Major damage results from larval feeding which leads to higher plant losses during winter, weakening of young plants, and facilitating the infestation of oilseed rape by fungal diseases and viruses (Godan, 1950; Bonnemaison, 1965; Schulz and Daebeler, 1984; Winfield, 1992).

The rape stem weevil starts to invade the oilseed rape fields on sunny days in early spring, when the maximum temperature reaches 10℃ (Schmutterer, 1956). After a short maturity feeding, females lay their eggs into the stems of oilseed rape plants close to the shoot tip (Dosse, 1951). Thereby they also excrete substances which cause abnormal growth of the plants and splitting of the stems (Lerin, 1993; Paul, 2003). The feeding wounds promote subsequent plant infestation by fungal pathogens (Broschewitz et al., 1993). The larvae feed inside the stems until they leave the plants to pupate in soil (Dosse, 1951). Adults persist in their pupation cave until the following spring (Günthart, 1949; Alford et al., 2003). The main damage is caused by oviposition and larval feeding inside stems (Lerin, 1995; Williams, 2010).

The cabbage stem weevils usually immigrate into the oilseed rape fields in March and April. They often occur together with the rape stem weevil (Günthart, 1949). After colonisation of their host plant the adults conduct a maturity feeding (Günthart, 1949). Subsequently they start to oviposite into petioles and rarely into stems of oilseed rape plants (Günthart, 1949). Neonate larvae feed inside the petioles before they move into stems (Günthart, 1949; Broschewitz and Daebeler, 1987). Here they complete their development before they leave the plants for pupation in soil (Günthart, 1949). New generation beetles emerge in summer (Günthart, 1949). After a short feeding period, the beetles leave the oilseed rape fields and migrate to their winter habitats (Günthart, 1949). Here they stay until crop colonisation in the following spring (Günthart, 1949). Economic yield losses are usually caused by larval feeding inside the plants (Kelm and Walczak, 1998; Kelm and Klukowski, 2000). This causes weakening of the plants and increases infestation by fungal diseases (Broschewitz et al., 1993).

Also in spring, when maximum temperature increases above 15°C, pollen beetles invade the oilseed rape fields (Fritzsche, 1957). Adults feed on pollen of a wide range of host plants but for oviposition they rely on brassicaceous species (Free and Williams, 1978). When beetles colonize oilseed rape crops before the plants start to flower, they feed on buds to get access to the pollen (Fritzsche, 1957; Williams and Free, 1978). Thereby they often damage the ovaries and induce abortion of these buds (Fritzsche, 1957; Williams and Free, 1978; Nilsson, 1988). After maturity feeding, females lay their eggs into the buds (Fritzsche, 1957). The larvae develop inside the buds and open flowers (Fritzsche, 1957; Williams and Free, 1978). During this time they feed on pollen without causing damage (Williams and Free, 1978). When the development of larvae is completed, they drop onto the ground to pupate in soil (Fritzsche, 1957). New generation adults emerge in summer and feed on pollen of Brassicaceae and other plants before they migrate to their winter habitats (Fritzsche, 1957). The feeding of adults on buds of oilseed rape often causes high economic losses (Fritzsche, 1957; Nilsson, 1987). Feeding of adults on open flowers or feeding of larvae seldom causes considerable damage (Fritzsche, 1957; Williams and Free, 1978).

The cabbage seed weevil invades the oilseed rape crops in early summer when the maximum temperature reaches 15°C (Edner and Daebeler, 1984). They feed on buds, flowers and pods of oilseed rape plants (Dmoch, 1965a; Williams and Free, 1978). Following maturity feeding females bite holes into young pods and lay one egg into the pod (Dmoch, 1965b; Edner and Daebeler, 1984). These pods are marked by a pheromone to avoid that other females use these pods for oviposition (Ferguson et al., 1999; Ferguson and Williams, 1991). Inside the pods each larva feeds three to five seeds during its development (Dmoch, 1965a; Edner and Daebeler, 1984). The mature larvae bite an exit hole into the wall of the pod and drop onto the soil for pupation (Dmoch, 1965a). New generation adults leave the oilseed rape fields and migrate to their winter habitats after a short period of feeding (Williams and Free, 1978). The main damage is caused by oviposition and feeding of larval exit holes (Dmoch, 1965a; Williams and Free, 1978; Edner and Daebeler, 1984; Williams, 2010). This can offer ports of entry for fungal pathogens and for oviposition of brassica pod midge (Williams, 2010).

The brassica pod midge usually invades the oilseed rape crops during flowering (Speyer, 1921; Fröhlich, 1956; Buhl, 1960). The midges are not able to fly over long distances, and they mainly infest the margins of fields (Kühne, 1967). For oviposition they are dependent on lesions in the pod wall because they are not able to penetrate older pods on their own (Buhl, 1957). The larvae suck inside the pods, thereby secreting salivary substances which cause deformation of pods and early shedding of the seeds (Speyer, 1921; Paul, 2003). The larvae fall onto the ground and pupate in soil (Fröhlich, 1956; Buhl, 1960). The brassica pod midge has usually three generations per year, but only the first generation most often causes economic yield losses in winter oilseed rape (Fröhlich, 1956; Buhl, 1960; Axelsen, 1992).

Pest management strategies

Today the management of insect pests on oilseed rape strongly relies on chemical insecticides, predominantely of the chemical group of pyrethroids (Williams, 2004). Due to adverse effects of these insecticides on non target insects, there is political pressure to reduce their usage (Thieme et al., 2010; Ulber et al., 2010; Walters et al., 2003). Recently, Bommarco et al. (2011) reported that the parasitism rate of diamond back moth was significantly reduced when insecticides were applied to cabbage crops. Moreover, insecticides in oilseed rape are often applied routinely, even though the pest infestation is below threshold level (Williams, 2004; 2010). In the past much effort has been spent to develop alternative strategies for pest management (Williams, 2004). One approach was to enhance the impact of natural enemies of pest insects (Walters et al., 2003). This can be achieved by reduced tillage after harvest of oilseed rape (Nilsson, 2010), which avoid destruction of larval parasitoids due to deep ploughing (Nitzsche and Ulber, 1998; Nilsson, 2010). The provision of flowering strips is another approach to promote these parasitoids, which then may help to regulate pest population (Büchi, 2002).

Another non-chemical strategy for management of insect pests in oilseed rape is the use of trap crops (Büchi, 1990; Cook, 2002; Nilsson, 2004; Barari et al., 2005; Büchs, 2009; Carcamo et al., 2007). Hokkannen (1991) defined trap crops as "plant stands that are grown to attract insects or other organisms like nematodes to protect target crops from pest attack. Protection may be achieved by preventing the pests from reaching the crop or by concentrating them in a certain part of the field". According to this, Shelton and Badenes-Perez (2006) differentiated between "conventional trap cropping" and "dead-end trap cropping". For conventional trap cropping a highly attractive trap crop is planted next to the targeted crop. Due to its high attractiveness, pest insects are maintained at the trap crop and do not colonize the main crop. In a "dead-end trap cropping" systems, the adult pests also prefer the attractive trap crop but in contrast to conventional trap cropping, the plants are not suitable as host for their offspring. This can either be achieved by insufficient host quality of the trap plant or by application of an insecticide. In order to protect oilseed rape from pest attack, most often turnip rape is used as a trap crop because it is preferred to oilseed rape by most specialist insect pests (Büchi, 1990; Lambdon et al., 1998; Cook et al., 2006; Cook et al., 2007). While the insects are maintained at the turnip rape borders, the main crop is protected against insect damage (Vandermeer, 1989; Hokkanen, 1991; Shelton and Badenes-Perez, 2006). Some studies have shown a positive effect of the trap crop strategy, however, the results of prior studies are inconsistent.

In earlier experiments two different types of experimental designs have been applied. Most often turnip rape was sown as a perimeter strip in order to trap the insect pests when they invade the field (Cook et al., 2004; Barari et al., 2005; Valantin-Morison, 2006; Carcamo et al., 2007). In the second design, a seed mixture of oilseed rape and turnip rape was sown to bound the pests to single turnip rape plants across the whole oilseed rape crop (Büchi, 1990; Nilsson, 2004). In both cases the proportion of the trap crop area in relation to the main crop area had an effect on the efficacy of the system (Hannunnen, 2005).

Previous studies also have shown that volatiles of essential oils emitted from non host plants have potential to distract pollen beetles from their host plants by masking the host plants' volatiles (Mauchline et al., 2005).

Host plant resistance can also provide a tool for integrated pest management. Sources of resistance against cabbage stem weevil and cabbage seed weevil recently were identified in rape cultivars and resyntheses (Eickermann et al., 2010; Tansey et al., 2011).

Host selection by herbivorous insects

Host selection by herbivorous insects is usually divided into three steps (Finch and Collier, 2000; Williams and Cook, 2010). The location of host plants is initiated by long distance orientation, which is commonly driven by olfactory cues (Finch and Collier, 2000). The cabbage seed weevil and other oilseed rape pests are attracted to degradation products of glucosinolates such as isothicyanates and nitriles as well as non-specific volatiles which are released from plants of various plant families (Bartlet et al., 1993; Bartlet et al., 1997; Cook et al., 2007). The orientation across intermediate distances is driven by visual stimuli such as colour and shape of host plants (Smart et al., 1997; Williams and Cook, 2010). Many specialist insects on oilseed rape are attracted to the yellow colour (Giamoustaris and Mithen, 1996; Smart et al., 1997). When the insect lands on a plants' surface, non-volatile substances are responsible for host plant acceptance (Finch and Collier, 2000).This prefeeding behaviour was investigated in detail in the crucifer flea beetle (*Phyllotreta cruciferae*), an important pest on canola in Canada (Lamb, 1989). When the insects attain a supposed host plant, three stages of behaviour were observed (Henderson et al., 2004). It starts with an acclimation phase, this is followed by stimulation and it ends up with initial feeding (Henderson et al., 2004). The insects perceive the stimuli from the host plant via chemical receptors at their tarsae and antennae and via test bites (Isidoro et al., 1998; Schoonhoven et al., 2005). For specialist insects on Brassicaceae, glucosinolates which are derived from secondary plant metabolism, are known to play a major role for host acceptance (Bartlet et al., 1994; Huang and Renwick, 1994). The feeding of cabbage stem flea beetles on a synthetic agar medium was significantly stimulated by gluconapin, glucotropaeolin, gluconasturtiin and glucobrassicin (Bartlet et al., 1994). Glucosinolates are known to be essential for host acceptance of this insect species (Bartlet and Williams, 1991; Williams and Bartlet, 1993). Further, secondary plant substances such as phenolics and alkaloids are also known to affect host acceptance and preference of other insect species (Mayer, 2004; Schoonhoven et al., 2005). Earlier studies supposed that cabbage stem flea beetles are also affected by plant chemicals which do not belong to the chemical class of glucosinolates (Bartlet & Williams, 1991). Moreover, sugars and amino acids which are produced in the primary metabolism of plants are known to have an effect on the feeding behaviour of herbivorous insects such as cabbage stem flea beetle (Bartlet et al., 1994; Vanloon & Vaneeuwijk, 1989). Some characteristics of leaf morphology are also known to affect insect-plant interactions. A high trichome density or a thick layer of epicuticular waxes, for example, are known to hamper flea beetles from feeding (Bodnaryk, 1992; Lamb, 1980; Lambdon et al., 1998; Soroka et al., 2011).

Objectives

A major objective of this study was to identify differences in susceptibility or even resistance of brassicaceous species to infestation by specialised insect pests. Furthermore, the mechanisms of resistance were investigated. This knowledge could be used in breeding programs to improve resistance against insect pests. To identify differences in host plant quality, the adult feeding preference and larval performance of cabbage stem flea beetle was assessed on a variety of brassicaceous species under laboratory conditions. Cabbage stem flea beetle was used as test insect in these experiments, because this species is a specialist on Brassicaceae and a major pest in German oilseed rape production. The plant species and lines which were used in these experiments were analysed for substances such as glucosinolates, sugars and phenolics. Trichome density on the leaf surface and dry weight of leaves were also measured. We tried to correlate these host plant characteristics to the feeding preference and larval performance, because these factors were assumed to play a major role in insect-plant interaction (see Schoonhoven, 2005).

Another objective was to assess the suitability of turnip rape as trap crop for control of insect pests on winter oilseed rape under on-farm conditions. Turnip rape was either grown in perimeter strips (Fig. 1) along the edges of oilseed rape fields or in mixtures containing 5% and 10% turnip rape, respectively. Moreover, in field plot experiments several turnip rape cultivars were tested for their suitability as trap crop. Therefore, cultivars with low erucic acid and low glucosinolate content in the seeds were compared to cultivars containing high contents of both substances in the seeds.

The overall goal was to develop strategies which could help to reduce insecticide use in oilseed rape production.

Fig 1: Turnip rape sown as a perimeter strip along the edges of an oilseed rape rape field; Ebergötzen 2011

References

Alford DV (2003) The oilseed rape crop. In: Biocontrol of oilseed rape pests (ed. DV Alford). Blackwell Science Ltd., Oxford, pp. 1-8.

Alford DV, Nilsson C & Ulber B (2003) Insect pests of oilseed rape crops. In: Biocontrol of oilseed rape pests (ed. DV Alford). Blackwell Science Ltd., Oxford, pp. 9-42.

Anonymus (2011) Bodennutzung und Ernte. Statistisches Bundesamt Deutschland.

Axelsen J (1992) The population dynamics and mortalities of the pod gall midge (*Dasyneura brassicae* Winn.) (Dipt., Cecidomyiidae) in winter rape and spring rape (*Brassica napus* L.) in Denmark. Journal of Applied Entomology 114: 463-471.

Barari H, Cook SM, Clark SJ & Williams IH (2005) Effect of a turnip rape (*Brassica rapa*) trap crop on stem-mining pests and their parasitoids in winter oilseed rape (*Brassica napus*). Biocontrol 50: 69-86.

Bartlet E, Blight MM, Hick AJ & Williams IH (1993) The responses of the cabbage seed weevil (*Ceutorhynchus assimilis*) to the odour of oilseed rape (*Brassica napus*) and to some volatile isothiocyanates. Entomologia Experimentalis et Applicata 68: 295-302.

Bartlet E, Blight MM, Lane P & Williams IH (1997) The responses of the cabbage seed weevil *Ceutorhynchus assimilis* to volatile compounds from oilseed rape in a linear track olfactometer. Entomologia Experimentalis et Applicata 85: 257-262.

Bartlet E, Parsons D, Williams IH & Clark SJ (1994) The influence of glucosinolates and sugars on feeding by the cabbage stem flea beetle, *Psylliodes chrysocephala*. Entomologia Experimentalis et Applicata 73: 77-83.

Bartlet E & Williams IH (1991) Factors restricting the feeding of the cabbage stem flea beetle (*Psylliodes chrysocephala*). Entomologia Experimentalis et Applicata 60: 233-238.

Bell JM (1995) Meal and by-product utilization in animal nutrition. In: Brassica oilseeds - production and utilization (ed. DS Kimber & DI McGregor). CAB International, Wallingford, pp. 301-338.

Bodnaryk RP (1992) Leaf epicuticular wax, an antixenotic factor in Brassicaceae that affects the rate and pattern of feeding of flea beetles, *Phyllotreta cruciferae* (Goeze). Canadian Journal of Plant Science 72: 1295-1303.

Bommarco R, Miranda F, Bylund H & Bjorkman C (2011) Insecticides suppress natural enemies and increase pest damage in cabbage. Journal of Economic Entomology 104: 782-791.

Bonnemaison L (1965) Insect pests of crucifers and their control. Annual Review of Entomology 10: 233-256.

Broschewitz B & Daebeler F (1987) Beitrag zur Biologie und Schadwirkung des Gefleckten Kohltriebrüßlers (*Ceutorhynchus quadridens* Panz.) am Winterraps. Nachrichtenblatt für den Pflanzenschutzdienst in der DDR 41: 34-37.

Broschewitz B, Steinbach P & Goltermann S (1993) Einfluss stengelbewohnender tierischer Schaderreger auf den Befall von Winterraps mit *Phoma lingam* und *Botrytis cinerea*. Gesunde Pflanzen 45: 106-110.

Büchi R (1990) Investigation on the use of turnip rape as a trap plant to control oilseed rape pests. IOBC/wprs Bulletin 13 (4): 32-39.

Büchi R (2002) Mortality of pollen beetle (*Meligethes* spp.) larvae due to predators and parasitoids in rape fields and the effect of conservation strips. Agriculture, Ecosystems & Environment 90: 255-263.

Büchs W (2009) Fangpflanzen (trap crop) als Methode zur Kontrolle tierischer Schädlinge im ökologischen Rapsanbau. Mitteilungen der Deutschen Gesellschaft für Allgemeine und Angewandte Entomologie 17: 241-246.

Buhl C (1957) Beitrag zur Frage der biologischen Abhängigkeit der Kohlschotenmücke (*Dasyneura brassicae* Winn.) von dem Kohlschotenrüßler (*Ceuthorrhynchus assimilis* Payk.). Zeitschrift für Pflanzenkrankheiten (Pflanzenpathologie) und Pflanzenschutz 64: 562-568.

Buhl C (1960) Beobachtungen über vermehrtes Schadauftreten der Kohlschotenmücke (*Dasyneura brassicae* Winn.) an Raps und Rübsen in Schleswig-Holstein. Nachrichtenblatt des deutschen Pflanzenschutzdienstes 12: 1-6.

Carcamo HA, Dunn R, Dosdall LM & Olfert O (2007) Managing cabbage seedpod weevil in canola using a trap crop - A commercial field scale study in western Canada. Crop Protection 26: 1325-1334.

Cook SM, L. E. Smart, R. J. P. Potting, E. Bartlet, J. L. Martin, D. A. Murray, N. P. Watts, I. H. Williams (2002) Turnip rape (*Brassica rapa*) as a trap crop to protect oilseed rape (*Brassica napus*) from infestation by insect pests: potential and mechanisms of action. The BCPC Conference: Pests & Diseases 2002 (2): 569-574.

Cook SM, Rasmussen HB, Birkett MA, Murray DA, Pye BJ, Watts NP & Williams IH (2007) Behavioural and chemical ecology underlying the success of turnip rape (*Brassica rapa*) trap crops in protecting oilseed rape (*Brassica napus*) from the pollen beetle (*Meligethes aeneus*). Arthropod-Plant Interactions 1: 57-67

Cook SM, Smart LE, Martin JL, Murray DA, Watts NP & Williams IH (2006) Exploitation of host plant preferences in pest management strategies for oilseed rape (Brassica napus). Entomologia Experimentalis et Applicata 119: 221-229.

Cook SM, Watts NP, Hunter F, Smart LE & Williams IH (2004) Effects of a turnip rape trap crop on the spatial distribution of *Meligethes aeneus* and *Ceutorhynchus assimilis* in oilseed rape. IOBC/wprs Bulletin 27 (10): 199-206.

Dmoch J (1965a) The dynamics of population of the cabbage seedpod weevil (*Ceutorhynchus assimilis* Payk.) and the development of winter rape. Part 1. Ekologia Polska Seria A 8: 249-287.

Dmoch J (1965b) The dynamics of population of the cabbage seedpod weevil (*Ceutorhynchus assimilis* Payk.) and the development of winter rape. Part 2. Ekologia Polska Seria A 8: 463-489.

Dosse G (1951) Der große Kohltriebrüssler *Ceutorrhynchus napi* (Gyll.) - Biologie, Schadauftreten und Bekämpfung unter besonderer Berücksichtigung der "Gallbildung" an Kohlpflanzen. Journal of Applied Entomology 32: 489-566.

Edner B & Daebeler F (1984) Zum Schadgeschehen durch den Kohlschotenrüssler (*Ceutorhynchus assimilis* Payk.). Nachrichtenblatt für den Pflanzenschutz in der DDR 38: 115-117.

Eickermann M, Ulber B & Vidal S (2010) Resynthesized lines and cultivars of *Brassica napus* L. provide sources of resistance to the cabbage stem weevil (*Ceutorhynchus pallidactylus* (Marsh.)). Bulletin of Entomological Research: 1-8.

Eurostat (2009) http://epp.eurostat.ec.europa.eu/portal/page/portal/ agriculture /data/database. date: 29.09.2011

Ferguson AW, Solinas M, Ziesmann J, Isidoro N, Williams IH, Scubla P, Mudd A, Clark SJ & Wadhams LJ (1999) Identification of the gland secreting oviposition-deterring pheromone in the cabbage seed weevil, *Ceutorhynchus assimilis*, and the mechanism of pheromone deposition. Journal of Insect Physiology 45: 687-699.

Ferguson AW & Williams IH (1991) Deposition and longevity of oviposition-deterring pheromone in the cabbage seed weevil. Physiological Entomology 16: 27-33.

Finch S & Collier RH (2000) Host-plant selection by insects - a theory based on 'appropriate/inappropriate landings' by pest insects of cruciferous plants. Entomologia Experimentalis et Applicata 96: 91-102.

Free JB & Williams IH (1978) The response of the pollen beetle, *Meligethes aeneus*, and the seed weevil, *Ceutorhynchus assimilis*, to oil-seed rape, *Brassica napus*, and other plants. Journal of Applied Ecology 15: 761-774

Fritzsche R (1957) Zur Biologie und Ökologie der Rapsschädlinge aus der Gattung *Meligethes*. Journal of Applied Entomology 40: 222-280.

Fröhlich G (1956) Zur Biologie und Bekämpfung der Kohlschotengallmücke (*Dasyneura brassicae* Winn). Nachrichtenblatt für den deutschen Pflanzenschutzdienst N.F. 10: 123-128.

Giamoustaris A & Mithen R (1996) The effect of flower colour and glucosinolates on the interaction between oilseed rape and pollen beetles. Entomologia Experimentalis et Applicata 80: 206-208.

Godan D (1950) Wann ist der Rapserdflohlarven-Befall für den Raspacker gefährlicher, im Herbst oder im Frühjahr? Nachrichtenblatt des deutschen Pflanzenschutzdienstes 2: 149-153.

Godan D (1951) Über Nahrungs- und Brutpflanzen des Rapserdflohs (*Psylliodes chrysocephala* L.). Anzeiger für Schädlingskunde 24: 81-84.

Günthart E (1949) Beiträge zur Lebensweise und Bekämpfung von *Ceutorrhynchus quadridens* PANZ. und *Ceutorrhynchus napi* GYLL. Mitteilungen der Schweizerischen Entomologischen Gesellschaft 22: 443-591.

Hannunnen S (2005) Modelling the interplay between pest movement and the physical design of trap crop systems. Agricultural and Forest Entomology 7: 11-20.

Henderson AE, Hallett RH & Soroka JJ (2004) Prefeeding behavior of the crucifer flea beetle, *Phyllotreta cruciferae*, on host and nonhost crucifers. Journal of Insect Behavior 17: 17-39.

Hokkanen HMT (1991) Trap cropping in pest management. Annual Review of Entomology 36: 119-138.

Huang XP & Renwick JAA (1994) Relative activities of glucosinolates as oviposition stimulants for *Pieris rapae* and *P. napi oleracea*. Journal of Chemical Ecology 20: 1025-1037.

Isidoro N, Bartlet E, Ziesmann J & Williams IH (1998) Antennal contact chemosensilla in *Psylliodes chrysocephala* responding to cruciferous allelochemicals. Physiological Entomology 23: 131-138.

Kaufmann O (1941) Zur Biologie des Rapserdflohes (*Psylliodes chrysocephala* L.). Zeitschrift für Pflanzenkrankheiten (Pflanzenpathologie) und Pflanzenschutz 51: 305-324.

Kelm M & Klukowski Z (2000) The effect of stem weevil (Ceutorhynchus pallidactylus Marsh.) infestation on oilseed rape yield. IOBC/wprs Bulletin 23 (6): 125-130.

Kelm M & Walczak B (1998) The relationship between the stem weevil (*Ceutorhynchus pallidactylus* Marsh.) injury and losses of flower buds. IOBC/wprs Bulletin 21 (5): 147-151.

Kimber DS & McGregor DI (1995) The species and their origin, cultivation and world production. In: Brassica oilseeds - production and utilization (eds. DS Kimber & DI McGregor). CAB International, Wallingford, pp. 1-8.

Kühne W (1967) Zur Verteilung der Kohlschotengallmücke *Dasyneura brassicae* Winnertz in großflächigen Rapsbeständen. Beiträge zur Entomologie 17: 287-297.

Lamb RJ (1980) Hairs protect pods of mustard (*Brassica hirta* Gisilba) from flea beetle feeding damage. Canadian Journal of Plant Science 60: 1439-1440.

Lamb RJ (1989) Entomology of oilseed *Brassica* crops. Annual Review of Entomology 34: 211-229.

Lambdon PW, Hassall M & Mithen R (1998) Feeding preferences of woodpigeons and flea-beetles for oilseed rape and turnip rape. Annals of Applied Biology 133: 313-328.

Lerin J (1993) Influence of the growth rate of oilseed rape on the splitting of the stem after an attack of *Ceutorhyncus napi* Gyll. IOBC/wprs Bulletin 16 (3): 160-163.

Lerin J (1995) Assessement of yield losses caused by insects in winter oilseed rape, a critical review. IOBC/wprs Bulletin 18 (4): 95-101.

Mauchline AL, Osborne JL, Martin AP, Poppy GM & Powell W (2005) The effects of non-host plant essential oil volatiles on the behaviour of the pollen beetle *Meligethes aeneus*. Entomologia Experimentalis et Applicata 114: 181-188.

Mayer AM (2004) Resistance to herbivores and fungal pathogens: Variations on a common theme? A review comparing the effect of secondary metabolites, induced and constitutive, on herbivores and fungal pathogens. Israel Journal of Plant Sciences 52: 279-292.

Nilsson C (1987) Yield losses in summer rape caused by pollen beetles (*Meligethes* spp.). Swedish Journal of Agricultural Research 17: 105-111.

Nilsson C (1988) Pollen beetles (Meligethes aeneus) and flowering in rape. Swedish Journal of Agricultural Research 18: 113-118.

Nilsson C (2004) Trap plants to avoid insecticide application against pollen beetles in oilseed rape. IOBC/wprs Bulletin 27 (10): 215-221.

Nilsson C (2010) Impact of soil tillage on parasitoids of oilseed rape pests. In: Biocontrol-based integrated management of oilseed rape pests (ed. IH Williams). Springer Science+Business Media B.V., Dordrecht, Heidelberg, London, New York, pp. 305-312.

Nitzsche O & Ulber B (1998) Influence of different tillage treatments following the harvest of oilseed-rape on the mortality of pollen beetle (*Meligethes* spp.) parasitoids. Journal of Plant Diseases and Protection 105: 417-421.

Paul VH (2003) Raps: Krankheiten, Schädlinge, Schadpflanzen. Verlag Th. Mann, Gelsenkirchen.

Saringer G (1984) Summer diapause of cabbage stem flea beetle, *Psylliodes chrysocephala* L. (Col. Chrysomelidae). Journal of Applied Entomology 98: 50-54.

Schmutterer H (1956) Zur Lebensweise und Bekämpfung des Großen Rapsstengelrüsslers *Ceutorrhynchus napi* (Gyll.). Journal of Applied Entomology 39: 302-315.

Schoonhoven LM, van Loon JJA & Dicke M (2005) Insect-plant biology. Oxford University Press, Oxford.

Schulz R-R (1992) Untersuchung zur Aktivität und zum Zuflugverhalten des Rapserdflohs (*Psylliodes chrysocephala* L.). Nachrichtenblatt des deutschen Pflanzenschutzdienstes 44: 6-8.

Schulz R-R & Daebeler F (1984) Zum Schaden durch den Rapserdfloh (*Psylliodes chrysocephala* L.), insbesondere seiner Imagines. Nachrichtenblatt für den Pflanzenschutz in der DDR 38: 113-115.

Shelton AM & Badenes-Perez E (2006) Concepts and applications of trap cropping in pest management. Annual Review of Entomology 51: 285-308.

Smart LE, Blight MM & Hick AJ (1997) Effect of visual cues and a mixture of isothiocyanates on trap capture of cabbage seed weevil, *Ceutorhynchus assimilis*. Journal of Chemical Ecology 23: 889-902.

Soroka JJ, Holowachuk JM, Gruber MY & Grenkow LF (2011) Feeding by flea beetles (Coleoptera: Chrysomelidae; *Phyllotreta* spp.) is decreased on canola (*Brassica napus*) seedlings with increased trichome density. Journal of Economic Entomology 104: 125-136.

Speyer W (1921) Beitrag zur Biologie der Kohlschotenmücke. Bericht über die Tätigkeit der Biologischen Reichsanstalt für Land- und Forstwirtschaft 16: 208-218.

Tansey JA, Dosdall LM & Keddie A (2011) Incorporation of novel *Ceutorhynchus obstrictus*-resistant canola genotypes into mixed cropping strategies: evidence for associational resistance. Journal of Insect Behavior 24: 116-131.

Thieme T, Heimbach U & Müller A (2010) Chemical control of insect pests and insecticide resistance in oilseed rape. In: Biocontrol-based integrated management of oilseed rape pests (ed. IH Williams). Springer Science+Business Media B.V., Dordrecht, Heidelberg, London, New York, pp. 313-336.

Ulber B, Klukowski Z & Williams IH (2010) Impact of insecticides on parasitoids of oilseed rape pests. In: Biocontrol-based integrated management of oilseed rape pests (ed. IH Williams). Springer Science+Business B.V., Dordrecht, Heidelberg, London, New York, pp. 337-356.

Uppström B (1995) Seed chemistry. In: Brassica oilseeds production and utilization (eds. DS Kimber & DI McGregor). CAB International, Wallingford, pp. 217-242.

Valantin-Morison M, L Quere (2006) Effects of turnip rape trap crops on oilseed rape pests in a network of organic farmers' fields. CD-ROM Proceedings of International Symposium on Integrated Pest Management in Oilseed Rape, 3.-5. April 2006, Göttingen, Germany

Vandermeer J (1989) The ecology of intercropping. Cambridge University Press, Cambridge. pp. 237

Vanloon JJA & Vaneeuwijk FA (1989) Chemoreception of amino-acids in larvae of 2 species of *Pieris*. Physiological Entomology 14: 459-469.

Walters KFA, Young JEB, Kromb B & Cox PD (2003) Management of oilseed rape pests. In: Biocontrol of oilseed rape pests (ed. DV Alford). Blackwell Science Ltd., Oxford, pp. 43-72.

Williams IH (2004) Advances in insect pest management of oilseed rape in Europe. In: Insect pest management (eds. AR Horowitz & I Ishaaya). Springer -Verlag, Berlin Heidelberg, pp. 181-208.

Williams IH (2010) The major insect pests of oilseed rape in Europe and their management: an overview. In: Biocontrol-based integrated management of oilseed rape pests (ed. IH Williams). Springer Science+Business B.V., Dordrecht, Heidelberg, London, New York, pp. 1-43.

Williams IH & Bartlet E (1993) Host plant factors affecting the feeding of cabbage stem flea beetle (*Psylliodes chrysocephala* L.) on oilseed rape. IOBC Bulletin 16 (9): 202-209.

Williams IH & Cook SM (2010) Crop location by oilseed rape pests and host location by their parasitoids. In: Biocontrol-based integrated management of oilseed rape pests (ed. IH Williams). Springer Science+Business Media B. V., Dordrecht Heidelberg London New York.

Williams IH & Free JB (1978) The feeding and mating behaviour of pollen beetles (*Meligethes aeneus* Fab.) and seed weevils (*Ceutorhynchus assimilis* Payk.) on oil-seed rape (*Brassica napus* L.). Journal of Agricultural Science 91: 453-459.

Williams JJW & Carden PW (1961) Cabbage stem flea beetle in East Anglia. Plant Pathology 10: 85-95.

Winfield AL (1992) Management of oilseed rape pests in Europe. Agricultural Zoology Reviews 5: 51-95.

Chapter II: Performance of cabbage stem flea beetle larvae (*Psylliodes chrysocephala* L.) on brassicaceous host plants with different glucosinolate profile

Abstract

The cabbage stem flea beetle (CSFB) (*Psylliodes chrysocephala* L.) (Col.: Chrysomelidae) is one of the most important insect pests in European winter oilseed rape production. Adult beetles feed on young leaves while the larvae feed within stems and petioles of various brassicaceous plant species. Larval infestation can result in significant crop damage. In this investigation, the host plant quality of four oilseed rape cultivars and seven other brassicaceous species for CSFB was assessed under controlled climate conditions. Larval weight and larval mortality was measured following 14 or 21 days of feeding inside the petioles of test plants. To determine the impact of secondary plant substances on performance of larvae, the glucosinolate (GSL) content was analysed in petioles of uninfested plants and infested plants 21 days after the start of larval infestation. Larval performance was not significantly different between the four oilseed rape (*Brassica napus* L.) cultivars tested but differed considerably between the other brassicaceous species. The weight of larvae was highest in turnip rape (*Brassica rapa* L. *silvestris*) and lowest in white mustard (*Sinapis alba* L.). The duration of larval development was increased in oilseed radish. The systemic GSL response in petioles to infestation by CSFB varied between different plant species. GSL content of current oilseed rape cultivars showed little difference between non-infested and infested plants while the aliphatic GSL group was increase in infested turnip rape cultivars. In contrast, in indian rape (*Brassica rapa* subsp. *dichotoma* Roxb.) the aliphatic and aromatic GSL decreased in infested plants. Larval weight was positively correlated with progoitrin and 4-hydroxyglucobrassicin. White mustard which provides inferior conditions for larval development has potential to introduce insect resistance into high yielding oilseed rape cultivars in breeding programs.

Introduction

Oilseed (*Brassica napus* L.) rape is the most important oil crop in central and northern Europe (Eurostat, 2009). During the whole growing period it is attacked by various specialist pest insects (Alford et al., 2003; Ekbom, 1995). In the humid regions of northern Europe, the cabbage stem flea beetle (CSFB) (*Psylliodes chrysocephala* L.) (Col.: Chrysomelidae) is among the most devastating pests of winter oilseed rape (Williams, 2010; Winfield, 1992). In some years it causes significant yield losses even at moderate levels of infestation (Nilsson, 1990). During autumn adult beetles feed externally on the leaves of young plants but economically important damage is caused by the larvae (Ekbom, 1995). Larvae feed within petioles and stems of young rapeseed plants, thereby increasing plant losses during winter (Ekbom, 1995; Williams, 2010). In addition, their feeding wounds can result in a higher incidence of fungal pathogens, such as stem canker (*Leptosphaeria maculans* Ces. & de Not.) (Schulz & Daebeler, 1984). Control of CSFB is usually achieved by insecticidal seed dressing or spray application of insecticides, most often by using synthetic pyrethroids (Williams, 2004). Due to the high selection pressure CSFB populations may develop resistance against this group of insecticides, as recently has been reported from northern Germany (Heimbach & Müller, 2011). Problems related to chemical control makes the research for alternative control strategies more important. The search for insect resistant cultivars or plant species may be a promising option, as already known for cabbage seed weevil (*Ceutorhynchus obstrictus* Mrsh.) and root fly (*Delia floralis* L.) which are also specialist pests of Brassicaceae (Birch, 1988; Carcamo et al., 2007; Eickermann et al., 2010). Such sources of resistance could be integrated in oilseed rape breeding programs. Insect resistant oilseed rape cultivars could also help to reduce the number of insecticide applications and their negative impact on the environment (Williams, 2010).

Many plant species are accepted as host plants by larvae and adults of CSFB. Most of these species belong to the family of Brassicaceae, but also tropaeolaceous and resedaceous plant species are accepted as host (Bartlet & Williams, 1991). All these plants have in common that they contain glucosinolates (GSL) (Bartlet et al., 1994), the predominant group of sulphur containing secondary plant substances in Brassicaceae (Mithen, 2001; Schoonhoven et al., 2005). Former studies have shown that the GSL profile varies between different brassicaceous species and plant organs (Sang et al., 1984). Due to their chemical similarities GSL can be subdivided in aliphatic, aromatic and indole compounds (Fahey et al., 2001; Hopkins et al., 2009). Even small changes in the side chain structure can have substantial influence on their biological activity (Mithen, 2001). By acting as feeding stimulants for specialist insects of the orders Coleoptera, Diptera and Lepidoptera GSL are known to play an important role in plant herbivore interactions (Hopkins et al., 2009). Furthermore, GSL are assumed to affect the larval performance of several insects specialized on Brassicaceae (Agrawal & Kurashige, 2003; Beekwilder et al., 2008; Mewis et al., 2008; Rohr et al., 2009). The aim of this investigation was to identify brassicaceous plant species and cultivars which hamper larval development of CSFB and to analyse the relationship between the GSL content in petioles and larval performance. Moreover, the effect of infestation by CSFB larvae on systemically induced changes of the GSL content and composition in different plant species was investigated. The GSL content was measured in petioles, the most important feeding site of CSFB larvae.

Materials and Methods

Plant Material

The test plants were selected to cover a wide range of different brassicaceous species and oilseed rape cultivars which differed largely in their glucosinolate profiles. Cultivars were chosen according to information from the literature or results of preliminary experiments. Four oilseed rape (*Brassica napus* L.) cultivars were included in the experiments. 'Robust', 'Grizzly' and 'Campala' contain low levels of GSL and erucic acid in their seeds (double low-quality) while seeds of 'Lembkes Normal' contain high levels of GSL and erucic acid. Three turnip rape (*Brassica rapa* L. *silvestris*) cultivars were selected: 'Largo' has a double low quality, whereas 'Malwira' and 'Perko' both contain high amounts of erucic acid and GSL in the seeds. Indian rape (*Brassica rapa* subsp. *dichotoma* Roxb.), white mustard (*Sinapis alba* L.) cultivar 'Accent', oilseed radish (*Raphanus sativa* L. var.

oleiformis Pers.) cultivar 'Adagio' and white cabbage (*Brassica oleracea* L. convar. *capitata* var. *alba*) cultivar 'Brunswijker' were also tested in the experiments. The plants were grown in a 3:1 mixture of potting soil (Fruhstorfer Erde Typ 25) and sand under glasshouse conditions. They were transferred to controlled climate conditions (18°C ±2°C, 65-85% rH and a day/night period of L16:D8 hours) one week before the experiments started. Plants were fertilized once a week with all essential nutrients by application of 50ml of a 2g/l "HakaPhos Blau®" solution.

Insects

New generation adults of CSFB emerging from soil after pupation were collected from crops of oilseed rape in July. They were maintained in plastic boxes (17,5cm * 13cm * 6cm) at 15°C during their summer aestivation. To provide food and moisture one oilseed rape leaf (cv Oase or Mozart) was offered to the beetles at least once a week. Following the end of aestivation in September, wet filter paper was placed into the boxes to stimulate egg laying of females (Williams et al., 2003). Eggs were collected weekly and stored in plastic boxes on wet sand at 2°C ± 1°C for several months. To induce hatching of larvae the eggs were transferred to 20°C ± 2°C one week before each experiment

Experimental setup

The experiments started five weeks after sowing when the plants were in the six to eight leaf stage. In all experiments the oilseed rape cultivar 'Robust' was chosen as a reference because of the low glucosinolate content in the leaves. In each experiment larval development on the reference cultivar was tested against the development on two other cultivars. Sixteen to twenty plants of each cultivar were randomly arranged in a climate chamber without contact to adjacent plants, to prevent larval migration between test plants. Neonate larvae were released on test plants 24 hours after hatching by using a fine brush. Five larvae were released on the petiole base of the fourth youngest leaf near the axilla of each plant. After inoculation the light was turned off for six hours. Plants were sprayed with water every hour to avoid drought stress of the larvae; relative humidity was near to 100%. The inoculated plants were kept at 18°C ±2°C, 65-85% rH and L16:D8 hours. Half of the plants was harvested 14 days and 21 days, respectively, after inoculation. The petioles and stems were dissected and larvae were recovered. Larval fresh weight was recorded using a microbalance (Sartorius MC 5) and head capsule width was measured as indicators of larval performance.

Glucosinolate analysis

The GSL content was analysed from petioles of both infested and uninfested plants. Samples of petioles were collected from five uninfested plants at the start of the experiment and from five infested plants 21 days after release of CSFB larvae. The third and fourth youngest petiole of five plants of each treatment was pooleded. The plant material was shock frozen at -80 °C immediately followed by freeze drying and grinding. The GSL were extracted by methanol; glucotropeolin was added as an internal standard in the analyses. The samples were treated with sulfatase and after incubation desulphoglucosinolates were analysed by HPLC as described by Cleemput & Becker (2011).

Statistical analysis

To compare mean weight of larvae in test plants and in the reference cultivar 'Robust' students t-test was used. Plants which did not contain any larvae were excluded from statistical analysis of weight and head capsule width. Correlations between the initial GSL content in the petioles and relative larval weight were analysed. No statistical analysis of the GSL content was conducted because no replicates were available. To compare the number of larvae recovered from plants, ANOVA was conducted followed by a post-hoc Tukey HSD test. For all statistical analyses Statistica 9.1 (StatSoft Inc.) was used.

Results

The development of larvae differed widely between the host plants tested (Fig. 1 & 2). Generally, the differences were more obvious at 14 days than at 21 days after infestation. Larval weight did not differ significantly between 'Robust' and the other oilseed rape cultivars, neither at 14 nor at 21 days of development within the plants. In turnip rape and indian rape the larval weight after 14 days was increased by 10 - 30% compared to 'Robust', however, these differences were not significant and were no longer present at 21 days after the start of experiment. After 14 days of feeding, the larval weight was significantly smaller on white mustard, oilseed radish or white cabbage than on 'Robust'.. The number of larvae recovered from plants after 14 days was significantly smaller and the larval mortality was higher only in white mustard as compared to 'Robust' (Tab. 1).

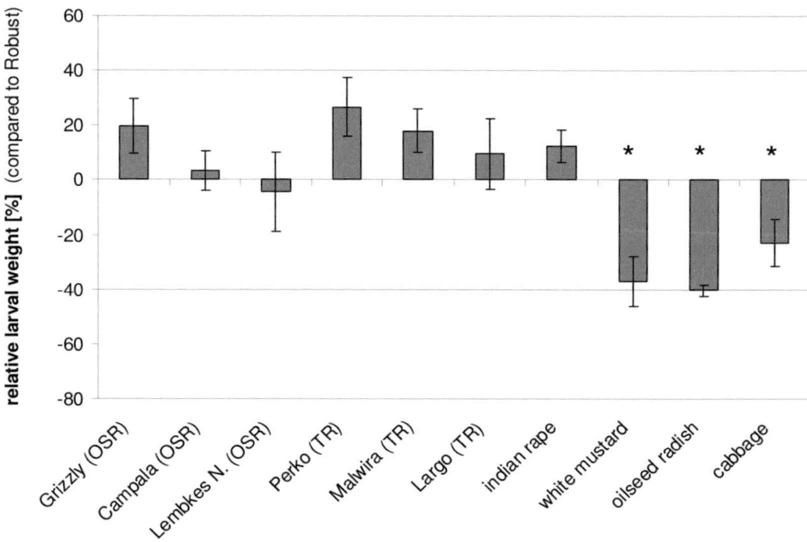

Fig. 1: Relative larval weight of *Psylliodes chrysocephala* reared in 10 test plants compared to larvae reared in the oilseed rape cultivar 'Robust' 14 days after release of larvae to plants. OSR - oilseed rape, TR - turnip rape; mean±SE; asterisks indicate significant differences to reference cultivar $\alpha \leq 0.05$; after 14 days larval fresh weight in 'Robust' varied from 1.7 to 2.4mg.

After 21 days significantly more larvae were found in oilseed radish because on 'Robust' and other plants tested a high proportion of larvae already had migrated to soil for pupation (Tab. 1). At 21 days after release the weight of larvae in oilseed radish was no longer smaller than the weight of larvae grown in 'Robust' while larvae feeding in white mustard and cabbage were still significantly lighter than larvae in 'Robust'.

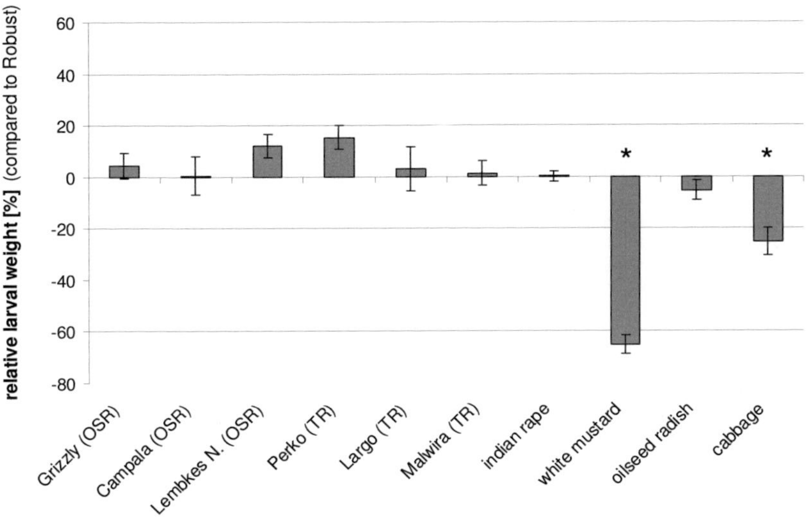

Fig. 2: Relative larval weight of *Psylliodes chrysocephala* reared in 10 test plants compared to larvae reared in the oilseed rape cultivar 'Robust' 21 days after release of larvae to plants. OSR - oilseed rape, TR - turnip rape; mean±SE; asterisks indicate significant differences to reference cultivar $\alpha \leq 0.05$; after 21 days larval fresh weight in 'Robust' varied from 4.5 to 5.3mg.

At 14 days after release the percentage of 3rd instar larvae was substantially lower in white mustard and oilseed radish than in 'Robust. Even at 21 days after release, 75% of larvae in white mustard had not developed to the 3rd larval instar (Tab. 1).

In plants not infested by larvae, the GSL content in petioles of the double low oilseed rape cultivars 'Robust', 'Grizzly' and 'Campala' was lower compared to 'Lembkes Normal' and other tested brassicaceous species (Tab. 2). In these three oilseed rape cultivars the GSL were mainly represented by the aliphatic fraction. The GSL in petioles of the old cultivar 'Lembkes Normal' was also dominated by the aliphatic fraction but the total content of GSL was 4.7 to 11 times higher than in the double low cultivars of oilseed rape. The aliphatic GSL were also predominant in the tested turnip rape cultivars but in contrast to oilseed rape in all turnip rape cultivars the aromatic GSL were on a higher level in the petioles.

Tab. 1: Number of larvae recovered 14 and 21 days after start of the experiments (dpi), proportion of dead larvae and percentage of larvae in third larval stage; different letters indicate significant differences between three cultivars included in each experimental run. Tukey HSD test α≤0.05

	14 dpi			**21 dpi**		
	larvae/plant	% dead	% 3rd instar	larvae/plant	% dead	% 3rd instar
Robust	4.4 a	0	89	1.5 a	6.2	93
Grizzly	4.5 a	0	93	1.4 a	0	100
Campala	4.0 a	7.5	90	2.0 a	0	100
Robust	2.1 a	0	75	1.5 a	0	100
Perko	2.9 a	0	61	1.1 a	0	100
Largo	2.5 a	0	60	2.3 a	0	100
Robust	4.3 a	0	93	0.7 a	0	100
Malwira	4.3 a	0	95	1.1 a	0	100
Ind. rape	4.2 a	0	98	1.0 a	0	100
Robust	4.1 a	2.4	98	1.1 a	0	91
Mustard	1.8 b	53.8	36	0.3 a	10	25
Radish	4.4 a	0	53	4.0 b	2.4	100
Robust	2.9 a	3.3	90	0.6 ab	25	100
Lembkes N.	2.0 a	9.1	62	0.2 a	66.7	67
Cabbage	2.7 a	3.6	82	1.6 b	11.1	100

Further, a high initial content of indole GSL was found particularly in 'Malwira' . Indian rape and white mustard both showed a high content of aromatic compounds whereas indole compounds were on a low level in both species. In contrast to all other plant species tested, the content of indole GSL and aromatic GSL in oilseed radish was on a relatively high level.

In petioles of white cabbage the total content of GSL was low and mainly represented by aliphatic compounds. While no correlation was detected between individual chemical groups of GSL and relative larval weight, positive correlations were found between relative larval weight after 14 days and initial content of progoitrin (df = 10; r = 0.65; p = 0.03) and 4-hydroxyglucobrassicin (df = 10; r = 0.63; p = 0.04) in petioles of host plants.

The composition of the GSL in the double low-quality oilseed rape cultivars did not show induced changes after 21 days of CSFB infestation. In contrast, the high initial content of ‘Lembkes Normal’ nearly doubled 21 days after larval infestation, resulting mostly from an increase in aliphatic GSL. All turnip rape cultivars showed a clear response to insect feeding with a strong increase of total GSL content. In contrast, in indian rape and white mustard the total GSL decreased following infestation by CSFB larvae.

Aromatic GSL compounds decreased in both species whereas aliphatics decreased in indian rape and increased in white mustard. Oilseed radish showed an increase in both aliphatic and aromatic GSL. In contrast the infestation by CSFB larvae had no effect on GSL content or composition in white cabbage cultivar ‘Brunswijker’.

Discussion

Our larval performance tests demonstrated that host plant quality of brassicaceous species for CSFB larvae is significantly different. While turnip rape cultivars were most suitable host plants white mustard clearly showed antibiosis resistance to the larvae. The high number of larvae present in oilseed radish after 21 days indicates that larval development was hampered in this host plant compared to oilseed rape. This was also confirmed by the low proportion of 3rd instar larvae after 14 days compared to larvae grown in ‘Robust’. In contrast to white mustard larval mortality was not increased in oilseed radish. The low number of larvae recovered from most test plants after 21 days compared to the number of larvae obtained after 14 days is due to the fact that a high proportion of the larvae had already completed their development and had left the plants for pupation in soil. Even though the larvae were treated the same way before each experimental run, the number of larvae found in the reference cultivar ‘Robust’ showed substantial differences between experimental runs. The factors responsible for the different mortality of larvae are not clear. Environmental and pre-treatment factors such as duration of storage of the eggs may have affected survival of larvae.

Tab. 2: Glucosinolate (GSL) content [µmol*g DM^{-1}] in petioles of uninfested (-) and infested (+) plants 21 days after inoculation; [1] progoitrin, sinigrin, gluconapoleiferin, glucoraphanin, gluconapin, glucobrassicanapin, glucoerucin; [2] 4-hydroxyglucobrassicin, glucobrassicin, 4-methoxyglucobrassicin, neoglucobrassicin; [3] sinalbin, gluconasturtiin

	Robust		Grizzly		Campala		Lembkes		Perko		Malwira		Largo		ind. rape		mustard		radish		cabbage	
	-	+	-	+	-	+	-	+	-	+	-	+	-	+	-	+	-	+	-	+	-	+
Aliphatic GSL [1]	0.9	0.9	0.4	0.4	0.4	1.3	6.0	10.8	7.6	11.8	3.7	9.3	9.1	27.7	1.7	0.9	0.3	1.9	0.4	0.4	1.1	1.1
Indole GSL [2]	0.2	0.2	0.2	0.1	0.1	0.3	0.2	0.9	0.7	1.0	1.9	0.8	0.6	2.0	0.3	0.2	0.1	0	3.7	5.0	0.8	0.6
Aromatic GSL [3]	0.3	0	0	0.2	0	0.3	0.4	0.6	2.5	2.5	0.7	1.1	1.1	2.0	16.3	11.7	6.6	4.5	2.0	3.8	0.3	0.4
total GSL	***1.4***	***1.1***	***0.6***	***0.7***	***0.5***	***1.9***	***6.6***	***12.3***	***10.8***	***15.3***	***6.3***	***11.2***	***10.8***	***31.7***	***18.3***	***12.8***	***7.0***	***6.4***	***6.1***	***9.2***	***2.2***	***2.1***

Earlier studies dealing with the performance of insects specialized on brassicaceous plants also have shown considerable differences in the host plant quality for larval development. For example, larvae of *Phaedon cochlearia* (F.) were heaviest on *Brassica rapa* and lightest on *Sinapis alba*, while the weight of larvae reared on *Raphanus sativus* was intermediate (Uddin et al., 2009). The ranking of host plant quality was similar to CSFB in our experiments; it has also been recorded for *Mamestra configurata* (McCloskey & Isman, 1993). Sarfraz et al. (2010) compared the host plant quality of swede (*Brassica napus* subsp. *rapifera*) and kale (*Brassica oleracea* var. *acephala*) for turnip root fly (*Delia floralis* Fallén) larvae and found that swede was a better host for larval growth. A similar difference between *B. napus* and *B. oleracea* was observed for CSFB in our experiments. An inferior host plant quality of *Sinapis alba* has been reported for cabbage seed weevil (*Ceutorhynchus obstrictus* Marsham). When seed weevil larvae were reared on pods of oilseed rape and white mustard, a significant reduction of larval weight was observed on white mustard (Dosdall & Kott, 2006; Tansey et al., 2010; Ulmer & Dosdall, 2006).

The factors which are responsible for the different host plant quality are not yet fully understood. For specialist insects on brassicaceous plants, like cabbage stem flea beetle or crucifer flea beetle (*Phyllotreta cruciferae* (Goeze)), secondary plant metabolites such as GSL can act as feeding stimulants (Bartlet & Williams, 1991; Feeny et al., 1970). The effect of GSL on larval performance was found to differ between insect species. The GSL content has a negative effect on larval development of generalist insects. For specialist insects, contradictory results on the effect of GSL on larval growth have been reported (Hopkins et al., 2009). Some authors found significant correlations between larval growth parameters and GSL content in plant tissue whereas others did not. Ulmer & Dosdall (2006) suggested that the aromatic GSL gluconasturtiin might be responsible for the delayed larval development and reduced larval weight of cabbage seedpod weevil in some brassicaceous species. However, in recent studies Tansey et al. (2010) found a positive correlation between larval weight of cabbage seed weevil and the content of gluconasturtiin in pods. The authors also reported a negative correlation between larval weight and the content of neoglucobrassicin in the pods. In experiments using *Arabidopsis thaliana* mutants, the larval performance of the two specialist insects *Pieris rapae* and *Pieris brassicae* was not affected by a drastic reduction of aliphatic or indole GSL in leaves (Mewis et al., 2008). No correlation between the GSL content and larval performance was found in studies on cabbage stem weevil (*Ceutorhynchus pallidactylus* Marsh.) and cabbage root fly (Birch et al., 1992; Eickermann et al., 2010). Bodnaryk (1997) concluded that resistance of white mustard against *Phyllotreta cruciferae* is independent from GSL content in the cotelydons. Our results suggest that the GSL are playing a role for host plant quality even for specialist insects such as CSFB. However, as the effect of single GSL on various insects was very heterogenous, further research is needed on this topic. Artificial diets or genetically manipulated host plants with modified GSL profiles may offer suitable tools for further experiments.

The effects of insect feeding on the GSL composition in plants have been repeatedly reported in literature (Koritsas et al., 1989; Koritsas et al., 1991; Birch et al., 1992; Bartlet et al., 1999), but results were not consistent. In our experiments the infestation by CSFB larvae had only little effect on indole GSL while aliphatic GSL increased in most test plants. In former studies feeding of adult CSFB on leaves of oilseed rape induced a systemic increase of indole GSL in young leaves seven days after feeding whereas the content of other GSL classes did not change significantly (Bartlet et al., 1999). Similar results were reported for non systemic responses of oilseed rape to infestation by CSFB larvae. Petioles infested by larvae showed a higher content of indole GSL and reduced content of aliphatic GSL compared to uninfested plants (Koritsas et al., 1989; 1991). When petioles of oilseed rape were mechanically damaged the GSL showed similar changes. In kale, indole as well as aliphatic GSL increased after mechanically damage whereas in white mustard only aromatic GSL increased (Koritsas et al., 1991). Contrasting observations were reported following infestation of swede, kale and oilseed rape by larvae of turnip root fly. Larvae feeding on the roots induced systemic changes of the GSL composition in aerial plant parts. The level of aliphatic GSL increased while indole GSL decreased in the leaves of these plants (Birch et al., 1992). Feeding by crucifer flea beetle had no effect on systemic induction of sinigrin and glucobrassicin (Traw & Dawson, 2002a; b).

Bartlet et al. (1999) suggested that a modified GSL profile may protect the plant against generalist insects and diseases while feeding by adult CSFB is not affected. A higher GSL content may attract higher numbers of specialist insects, because degradation products of GSL, isothiocyanates, are used for host finding (Bartlet, 1992). For future experiments on this topic 'Lembkes Normal' and 'Largo' are promising cultivars because of their very strong response in their GSL content compared to other brassicaceous species in our experiments. Field experiments need to be conducted to evaluate the performance of larvae of cabbage stem flea beetle on different host plant species under natural conditions. In the field the extended larval development may also increase the parasitation rate because larvae are exposed to their enemies for a longer time (Singh & Singh, 2005). Plants which hamper the larval development of cabbage stem flea beetle larvae may also increase efficacy of natural control mechanisms (Singh & Singh, 2005). Furthermore, other plant substances such as the fiber content in the petioles of different host plants have to be considered, due to the fact that the digestibility of plant material affects the larval performance (Waldbauer, 1964).

References

Agrawal AA & Kurashige NS (2003) A role for isothiocyanates in plant resistance against the specialist herbivore Pieris rapae. Journal of Chemical Ecology 29: 1403-1415.

Alford DV, Nilsson C & Ulber B (2003) Insect pests of oilseed rape crops: Biocontrol of oilseed rape pests (ed. DV Alford) Blackwell Science Ltd., Oxford, pp. 9-42.

Bartlet E, I. H. Williams, M. M. Blight, A. J. Hick (1992) Response of the oilseed rape pests, *Ceutorhynchus assimilis* and *Psylliodes chrysocephala*, to a mixture of isothiocyanates. Proceedings of the 8th International Symposium on Insect-Plant Relationships: 103-104.

Bartlet E, Kiddle G, Williams I & Wallsgrove R (1999) Wound-induced increases in the glucosinolate content of oilseed rape and their effect on subsequent herbivory by a crucifer specialist. Entomologia Experimentalis et Applicata 91: 163-167.

Bartlet E, Parsons D, Williams IH & Clark SJ (1994) The influence of glucosinolates and sugars on feeding by the cababge stem flea beetle, *Psylliodes chrysocephala*. Entomologia Experimentalis et Applicata 73: 77-83.

Bartlet E & Williams IH (1991) Factors restricting the feeding of the cabbage stem flea beetle (*Psylliodes chrysocephala*). Entomologia Experimentalis et Applicata 60: 233-238.

Beekwilder J, van Leeuwen W, van Dam NM, Bertossi M, Grandi V, Mizzi L, Soloviev M, Szabados L, Molthoff JW, Schipper B, Verbocht H, de Vos RCH, Morandini P, Aarts MGM & Bovy A (2008) The impact of the absence of aliphatic glucosinolates on insect herbivory in *Arabidopsis*. Plos One 3: 12.

Birch ANE (1988) Field and glasshouse studies on the components of resistance to root fly attack in swedes. Annals of Applied Biology 113: 89-100.

Birch ANE, Griffiths DW, Hopkins RJ, Smith WHM & McKinlay RG (1992) Glucosinolate responses of swede, kale, forage and oilseed rape to root damage by turnip root fly (*Delia floralis*) larvae. Journal of the Science of Food and Agriculture 60: 1-9.

Bodnaryk RP (1997) Will low-glucosinolate cultivars of the mustards *Brassica juncea* and *Sinapis alba* be vulnerable to insect pests? Canadian Journal of Plant Science 77: 283-287.

Carcamo H, Olfert O, Dosdall L, Herle C, Beres B & Soroka J (2007) Resistance to cabbage seedpod weevil among selected Brassicaceae germplasm. Canadian Entomologist 139: 658-669.

Cleemput S & Becker HC (2011) Genetic variation in leaf and stem glucosinolates in resynthesized lines of winter rapeseed (*Brassica napus* L.). Genetic Resources and Crop Evolution. doi:10.1007/s10722-011-9701-x.

Dosdall LM & Kott LS (2006) Introgression of resistance to cabbage seedpod weevil to canola from yellow mustard. Crop Science 46: 2437-2445.

Eickermann M, Ulber B & Vidal S (2010) Resynthesized lines and cultivars of *Brassica napus* L. provide sources of resistance to the cabbage stem weevil (*Ceutorhynchus pallidactylus* (Marsh.)). Bulletin of Entomological Research 110: 1-8.

Ekbom B (1995) Insect pests: Brassica oilseeds - production and utilization (ed. by DS Kimber & DI McGregor) CAB International, Wallingford, pp. 141-152.

Eurostat (2009) http://epp.eurostat.ec.europa.eu/portal/page/portal/agriculture/data/database. date: 29.09.2011

Fahey JW, Zalcmann AT & Talalay P (2001) The chemical diversity and distribution of glucosinolates and isothiocyanates among plants. Phytochemistry 56: 5-51.

Feeny P, Paauwe KL & Demong NJ (1970) Flea beetles and mustard oils: host plant specifity of *Phyllotreta cruciferae* and *P. striolata* adults (Coleoptera: Chrysomelidae). Annals of the Entomological Society of America 63: 832-841.

Heimbach U & Müller A (2011) Pyrethroid resistance of oilseed rape pest insects from 2005-2010 and German insecticide resistance management strategy: 13th International Rapeseed Congress, Prague, pp. 1278-1281.

Hopkins RJ, van Dam NM & van Loon JJA (2009) Role of glucosinolates in insect-plant relationships and multitrophic interactions. Annual Review of Entomology 54: 57-83.

Koritsas VM, Lewis JA & Fenwick GR (1989) Accumulation of indole glucosinolates in *Psylliodes chrysocephala* L. infested, or *Psylliodes chrysocephala* L. damaged tissues of oilseed rape (*Brassica napus* L.). Experientia 45: 493-495.

Koritsas VM, Lewis JA & Fenwick GR (1991) Glucosinolate response of oilseed rape, mustard and kale to mechanical wounding and infestation by cabbage stem beetle (*Psylliodes chrysocephala*). Annals of Applied Biology 118: 209-221.

McCloskey C & Isman MB (1993) Influence if foliar glucosinolates in oilseed rape and mustard on feeding and growth of bertha armyworm, *Mamestra configurata* Walker. Journal of Chemical Ecology 19: 249-266.

Mewis I, Rohr F, Tokuhisa JG & Ulrichs C (2008) Auswirkung einer Modifizierung des Glucosinolatprofils von *Arabidopsis thaliana* (L.) auf verschieden spezialisierte Lepidoptera. Mitteilungen der Deutschen Gesellschaft für Allgemeine und Angewandte Entomologie 16: 167-169.

Mithen R (2001) Glucosinolates - biochemistry, genetics and biological activity. Plant Growth Regulation 34: 91-103.

Nilsson C (1990) Yield losses in winter rape caused by cabbage stem flea beetle larvae (*Psylliodes chrysocephala* (L.)). IOBC/wprs Bulletin 13 (4): 53-55.

Rohr F, Ulrichs C & Mewis I (2009) Variability of aliphatic glucosinolates in *Arabidopsis thaliana* (L.) - Impact on glucosinolate profile and insect resistance. Journal of Applied Botany and Food Quality 82: 131-135.

Sang JP, Minchinton IR, Johnstone PK & Truscott RJW (1984) Glucosinolate profiles in the seed, root and leaf tissue of cabbage, mustard, rapeseed, radish and swede. Canadian Journal of Plant Science 64: 77-93.

Sarfraz RM, Dosdall LM & Keddie BA (2010) Performance of the specialist herbivore *Plutella xylostella* (Lepidoptera: Plutellidae) on Brassicaceae and non-Brassicaceae species. Canadian Entomologist 142: 24-35.

Schoonhoven LM, van Loon JJA & Dicke M (2005) Insect-plant biology. Oxford University Press, Oxford.

Schulz R-R & Daebeler F (1984) Zum Schaden durch den Rapserdfloh (*Psylliodes chrysocephala* L.), insbesondere seiner Imagines. Nachrichtenblatt für den Pflanzenschutz in der DDR 38: 113-115.

Singh DP & Singh A (2005) Disease and pest resistance in plants. Science Publishers, Enfield.

Tansey JA, Dosdall LM, Keddie A, Fletcher RS & Kott LS (2010) Antixenosis and antibiosis resistance to *Ceutorhynchus obstrictus* in novel germplasm derived from *Sinapis alba* x *Brassica napus*. Canadian Entomologist 142: 212-221.

Traw MB & Dawson TE (2002a) Differential induction of trichomes by three herbivores of black mustard. Oecologia 131: 526-532.

Traw MB & Dawson TE (2002b) Reduced performance of two specialist herbivores (Lepidoptera: Pieridae, Coleoptera: Chrysomelidae) on new leaves of damaged black mustard plants. Environmental Entomology 31: 714-722.

Uddin MM, Ulrichs C, Tokuhisa JG & Mewis I (2009) Impact of glucosinolate structure on the performance of the crucifer pest *Phaedon cochleariae* (F.). Journal of Applied Botany and Food Quality 82: 108-113.

Ulmer BJ & Dosdall LM (2006) Glucosinolate profile and oviposition behavior in relation to the susceptibilities of Brassicaceae to the cabbage seedpod weevil. Entomologia Experimentalis et Applicata 121: 203-213.

Waldbauer GP (1964) The consumption, digestion and utilization of solanaceous and non-solanaceous plants by larvae of the tobacco hornworm, *Protoparce sexta* (Johan.) (Lepidoptera: Sphingidae). Entomologia Experimentalis et Applicata 7: 253-269.

Williams IH (2004) Advances in insect pest management of oilseed rape in Europe. In: Insect pest management (eds. AR Horowitz & I Ishaaya). Springer-Verlag, Berlin Heidelberg, pp. 181-208.

Williams IH (2010) The major insect pests of oilseed rape in Europe and their management: an overview. In: Biocontrol-based integrated management of oilseed rape pests (ed. IH Williams). Springer Science+Business B.V., Dordrecht, Heidelberg, London, New York, pp. 1-43.

Williams IH, Büchi R & Ulber B (2003) Sampling, trapping and rearing oilseed rape pests and their parasitoids. In: Biocontrol of oilseed rape pests (ed. DV Alford). Blackwell Science Ltd, Oxford, pp. 145-160.

Winfield AL (1992) Management of oilseed rape pests in Europe. Agricultural Zoology Reviews 5: 51-95.

Chapter III: Effects of host plant characteristics on the feeding preference of adult cabbage stem flea beetle (*Psylliodes chrysocephala* L.)

Abstract

The cabbage stem flea beetle (CSFB) (*Psylliodes chrysocephala* L.) (Col. Chrysomelidae) is a major pests on oilseed rape in northern and central Europe. Present control strongly relies on application of insecticides. As an alternative strategy cultivars are needed which provide resistance against this insect pests. In this investigation various cultivars of turnip rape (*Brassica rapa*), oilseed rape (*Brassica napus*) and other brassicaceous species (*Sinapis alba*, *Raphanus sativa*, *Brassica oleracea* var. *capitata*) were screened for resistance to CSFB. A novel dual choice test was developed for assessing the feeding preference of adult CSFB. Leaves of a test plant were supplied simultaneously with leaves of a reference oilseed rape cultivar and to adult beetles. The leaf area consumed by CSFB was measured after 48 h. In addition, the content of glucosinolates, sugars and phenolics in the leaf lamina as well as the trichome density and specific dry weight of the tested leaves were analysed. The beetles significantly preferred cultivars of turnip rape and oilseed radish to oilseed rape. Oilseed rape was preferred when offered together with white cabbage. Adult CSFB showed no feeding preference between the tested oilseed rape cultivars. The content of glucosinolates and sugars as well as the dry matter content of the leaves showed considerable differences between the tested plant species while the total content of phenolics did not differ. A significant correlation was observed between the leaf area consumed by adults and trichome density. No correlation was observed between the feeding preference and other plant factors studied, such as content of different glucosinolates. As white cabbage was least preferred, it might play a role in future breeding programs.

Introduction

Oilseed rape (*Brassica napus* L.) crops in Europe are host to a high number of insect herbivores but only a few of them regularly cause significant losses (Alford et al., 2003). Cabbage stem flea beetle (CSFB) (*Psylliodes chrysocephala* L.) (Col.

Chrysomelidae) is a major insect pest which can cause considerable economic losses in some years is (Nilsson, 1990). Following aestivation, in late summer adult beetles invade crops of oilseed rape to feed on leaves of the young plants (Williams, 2010). Similar to other insect pests on oilseed rape, control of CSFB relies on the application of chemical insecticides. In autumn insecticidal seed treatment and spray applications of synthetic pyrethroids are applied for control of adults as well as larvae of CSFB (Williams, 2004). This may cause a high selective pressure on CSFB populations and increase the probability of resistance development, as has been recently reported for regional populations in northern Germany (Heimbach and Müller, 2011). Further, the excessive use of insecticides may have negative effects on the environment and suppress beneficial insects in agro-ecosystems (Williams, 2004; Bommarco et al., 2011). These arguments and the political pressure to reduce the risks of pesticide usage increase the need for implementation of an integrated pest management strategy. One important approach is the search for insect resistant cultivars (Evans and Scarisbrick, 1994). Resistance against some other oilseed rape pests has been reported in the literature (Birch, 1988; Carcamo et al., 2007; Eickermann et al., 2010; Tansey et al., 2010).

Adults of CSFB feed on a wide range of glucosinolate-containing plants. Nearly all reported host plants belong to the family of Brassicaceae (Bartlet and Williams, 1991). Within their range of host plants the beetles show clear preferences for some brassicaceous species, especially for turnip rape (*Brassica rapa* L.) (Büchi, 1995; Lambdon et al., 1998; Barari et al., 2005). The factors which are responsible for discrimination between host plants by CSFB are not yet fully understood. Some sugars and glucosinolates (GSL) are known to act as feeding stimulants for CSFB (Williams and Bartlet, 1993; Bartlet et al., 1994). In addition, amino acids and phenolics originating from the primary and secondary metabolism of plants are known to act as feeding stimulants or deterrents for other insect species (Bernays, 1981; Onyilagha et al., 2004; Schoonhoven et al., 2005). Furthermore, morphological leaf characters such as trichome density or trichome length are known to affect feeding by Chrysomelid beetles (Hoxie et al., 1975; Schoonhoven et al., 2005). Most of the previous choice experiments on CSFB were conducted by using artificial media or by supplying excised leaves (Bartlet and Williams,

1991; Bartlet et al., 1994). In this study a novel biotest was developed for dual choice feeding tests on intact plants. By using this method, various brassicaceous cultivars or species were screened for their susceptibility or partial resistance to CSFB feeding. Furthermore the impact of GSL, phenolics and sugar content of the leaves as well as leaf morphological cues on feeding preference was analysed in order to identify key factors which determine host plant attractiveness.

Materials and Methods

Insects and plant material

Pre-diapause adults of CSFB were captured from volunteer oilseed rape plants near Göttingen (Germany) at the end of June. During their summer aestivation the beetles were kept in plastic boxes (17,5cm * 13cm * 6cm) at 15°C. They had free access to oilseed rape leaves (cv. Oase or Mozard) which were renewed once a week. The bottom of the plastic box was covered with moistened filter paper to supply water. The completion of aestivation was indicated by an increased feeding activity of the beetles. Before used in the experiments, beetles were starved for 48 hours and sexes were separated.

The selection of brassicaceous species and cultivars for the experiments was based on results of earlier studies and reports in the literature (Tab. 1). Plants for dual choice tests and plants for analyses of leaf chemicals and leaf morphology were grown together in a greenhouse until the four true leaf stage. The substrate consisted of 75% potting soil (Fruhstorfer Erde Typ 25) and 25% sand. Subsequently, plants were kept in a climate chamber at 20°C ±2°C, 65-85% relative humidity and a light regime of 16L:8D until the six to eight true leaf stage. The plants were fertilized once a week with 50ml of 0.2% HakaPhos Blau® (15%N, 10%P_2O_5, 15%K_2O, 2%MgO, 0.01%B, 0.02%Cu, 0.075%Fe, 0.05%Mn, 0.001%Mo, 0.015%Zn) solution to supply all essential nutrients.

Tab. 1: Set of plants included in dual choice experiments, leaf substance analyses and leaf morphology assessment

Cultivar name	Common name	Botanical name
Robust	Oilseed rape	*Brassica napus* L
Emerald	Fodder rape	*Brassica napus* L.
Grizzly	Oilseed rape	*Brassica napus* L.
Lembkes Normal	Oilseed rape	*Brassica napus* L.
Campala	Oilseed rape	*Brassica napus* L.
Largo	Turnip rape	*Brassica rapa* L. *silvestris* Briggs
Perko PVH	Turnip rape	*Brassica rapa* L. *silvestris* Briggs
Malwira	Turnip rape	*Brassica rapa* L. *silvestris* Briggs
Salut	Turnip rape	*Brassica rapa* L. *silvestris* Briggs
Accent	White mustard	*Sinapis alba* L.
Adagio	Oilseed radish	*Raphanus sativa* L. var. *oleiformis* Pers.
Brunswijker	White cabbage	Brassica oleracea L. var. capitata f. alba

Dual choice clip cages

Clip cages as described by Adams and van Emden (1972) were modified for the dual choice experiments (Fig. 1). A cylinder of Perspex (diameter 6cm, height 1.5cm) was covered by lids of Petri dishes (diameter 6cm) on both opposite sites. Two leaves, one of each test plant, were inserted between the cylinder and the Petri dish. The leaves were aligned with their upper surface facing to the cylinder. The edges of the cylinder and the Petri dishes were lined with foamed plastic to protect the leaf surface from damage. The cage was hold together by using a hair clip and was supported by a wooden stick to align leaves in a vertical position. In all dual choice experiments the oilseed rape cultivar 'Robust' was used as a reference cultivar because of its low GSL content in the leaf lamina as compared to other oilseed rape cultivars. To avoid variable effects of leaf age on the feeding behaviour only the 3rd youngest leaf was used for the dual choice tests. In each of 20 replicated tests, one leaf of cultivar 'Robust' was offered together with one leaf of the test cultivar. One female and one male of CSFB were released into the cage and allowed to feed on the upper leaf surfaces for 24 hours. After removal of the cages, the area consumed by the two beetles from each leaf was measured by using an image analysing software (Digitrace V 2.28). The feeding preference index (Q-value) was calculated by dividing the leaf area consumed from the test cultivar by the sum of consumed leaf areas of the reference and test cultivar (Larsen, 1985).

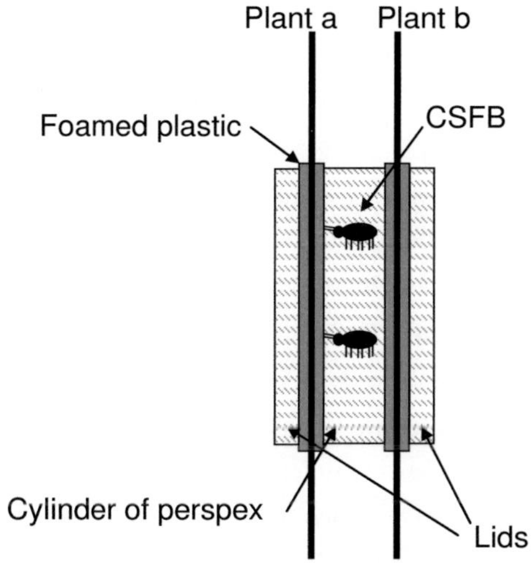

Fig. 1: Drawing of a dual choice clip cage attached to the leaves of two intact plants (plant a = reference cultivar and plant b = test cultivar); the central cylinder gives adult cabbage stem flea beetles (CSFB) access to eigher leaves.

Morphological leaf characteristics

When the plants were in the six to eight true leaf stage, leaf trichomes of the third youngest leaf were counted. Five leaf discs of 1cm² each were cut using a cork borer from the tip of each leaf, at the same position where the double clip cages were attached in dual choice tests,. The number of trichomes per cm² was determined from the upper site of the leaf discs under a microscope.

The dry weight of the third youngest leaf was assessed in the six to eight true leaf stage. Leaf discs of defined area of 12.6cm² were cut by using a cork borer. They were dried for 24 hours at 60°C until they reached a constant weight and weighed again. The dry weight was calculated as dry weight per cm² leaf area. Trichome density and dry weight were assessed from one leaf of five plants each.

Analysis of glucosinolates, sugars and phenolics

Leaf samples for analyses of glucosinolates, sugars and phenolics were taken from plants of the same age as plants used for dual choice experiments. The third and fourth youngest leaf was collected from five plants of each tested plant species. Only the leaf laminas were used for the analyses. Samples of ten leaves were combined and immediately shock frozen at -80°C. The samples were freeze dried and ground with a rotor mill using a mesh size of 1mm.

For the GSL analyses, 200mg of the plant material was incubated two times at 75°C with 70% and 10% methanol solution, respectively. Glucotropeolin was added as internal standard. The supernatant was collected after each incubation and the combined GSL extract was transferred to a Sephadex column. The GSL were degraded by adding sulphatase at 39°C and the desulphoglucosinolates were analysed by using HPLC (Cleemput and Becker, 2011).

The sugar content in the leaves was analysed by HPLC as well (Keutgen and Pawelzik, 2008). The sugars were extracted from 500mg dried leaf material by incubating with 10ml demineralised water for 1 hour at 60°C in a water bath. After incubation, Carrez I and II solution were added for protein precipitation and the supernatant was purified by centrifugation and filtration. The leaf extracts were stored at -18°C until they were analysed.

The phenolics in the leaf lamina were measured photometrically as described by Singleton and Rossi (1965). Two subsamples of 250mg ground leaf material were extracted by using cooled 80% ethanol. The photometer was calibrated with different concentrations of a gallic acid solution (11.94 to 59.7 µg/ml) at a wavelength of 735.8nm just before the samples were analysed.

Content of sugar and phenolics of ‘Lembkes Normal’ could be determined.

Data analysis

The leaf area consumed from the reference cultivar and the test plant was compared using Mann-Whitney U-Test. To analyse the differences between trichome densities and dry weights of the leaves Kruskal-Wallis-ANOVA or ANOVA was conducted, respectively. The ANOVA was followed by post-hoc

Tukey HSD test. To determine the relationship between the feeding preference (Q-value) of CSFB and chemical plant substances or morphology correlation analyses were conducted. Differences were regarded as significant at $p \leq 0.05$. For all statistical analyses Statistica 9.1 (StatSoft Inc.) was used. No statistical analyses were conducted for the contents of glucosinolates, phenolics and sugars because no replicated datasets were available.

Results

The feeding activity of adult CSFB differed considerably between different species of host plants (Tab. 2). The leaf area consumed from the five cultivars of oilseed rape was not significantly different, as indicated by the dual choice test of ‘Robust’ with ‘Campala’, ‘Grizzly’, ‘Lembkes Normal’ and ‘Emerald’. When beetles were given a choice between oilseed rape and turnip rape, turnip rape was preferred in every combination. There was no significant difference between the tested cultivars of turnip rape. When oilseed rape and white mustard were offered, no clear preference for one of the plant species could be identified. Leaves of oilseed radish were preferred significantly to oilseed rape while leaves of white cabbage were less preferred than oilseed rape. However, the differences between the leaf area consumed in this test were not statistically significant.

The content and composition of GSL in the leaves showed a wide variation between the tested cultivars and species (Tab. 3). The GSL content in leaves of the oilseed rape cultivar ‘Robust’ was at a similar level in the three experimental runs, but it showed clear differences compared to the other oilseed rape cultivars. The total GSL content in ‘Robust’ was lower than in all other oilseed rape cultivars. While aliphatic GSL predominated, only small amounts of indole and aromatic GSL were detected.

Similarly, the content of GSL in the tested turnip rape cultivars was also dominated by aliphatic substances while aromatic or indole GSL played a minor role. The total GSL contents in the leaf lamina of the turnip rape cover crop cultivars ‘Perko PVH’ and ‘Malwira’ were much higher than in the turnip rape cultivars ‘Largo’ and ‘Salut’ (double low seed quality).

Tab. 2: Mean (±SE) leaf area (cm²) consumed by *P. chrysocephala* from the reference oilseed rape cv. 'Robust' and the test cultivar; dual choice feeding experiments; Q-value according to Larsen et al. 1985); p-value (U-test)

Test cultivar	*Species*	*Consumed leaf area [cm²]*		*Q-value*	*p-value (U-test)*
		reference plant	*test plant*		
Perko PVH	*B. rapa*	0.02 (±0.01)	0.23 (±0.05)	**0.92**	>0.001
Malwira	*B. rapa*	0.01 (±0.00)	0.11 (±0.02)	**0.91**	0.014
Salut	*B. rapa*	0.03 (±0.01)	0.18 (±0.02)	**0.87**	>0.001
Largo	*B. rapa*	0.02 (±0.01)	0.08 (±0.02)	**0.81**	>0.001
Adagio	*R. sativa*	0.05 (±0.02)	0.13 (±0.03)	**0.73**	0.014
Grizzly	*B. napus*	0.03 (±0.01)	0.05 (±0.02)	**0.65**	0.519
Campala	*B. napus*	0.04 (±0.02)	0.05 (±0.02)	**0.54**	1.000
Accent	*S. alba*	0.10 (±0.03)	0.08 (±0.03)	**0.45**	0.352
Lembkes N.	*B. napus*	0.06 (±0.02)	0.04 (±0.01)	**0.41**	0.612
Emerald	*B. napus*	0.07 (±0.01)	0.05 (±0.01)	**0.39**	0.057
Brunswijker	*B. oleracea*	0.04 (±0.02)	0.02 (±0.01)	**0.32**	0.889

The highest total content of GSL was found in the leaf lamina of white mustard, with the aromatic GSL sinalbin as a key component. The GSL of oilseed radish mainly consisted of indole and aromatic compounds whereas the total GSL content was intermediate compared to other plant species. As observed in oilseed rape and turnip rape, aliphatic GSL also predominated in white cabbage but leaves of white cabbage also contained relatively high amounts of indole compounds.

Tab. 3: Content of glucosinolates in leaf lamina of test plants (µmol/g DM), [1] aliphatic glucosinolates, [2] indolyl glucosinolates, [3] aromatic glucosinolates (PRO = progoitrin, SIN = sinigrin, GNL = gluconapoleiferin, RAA = glucoraphanin, SIB = sinalbin, GNA = gluconapin, 4OH = 4-hydroxyglucobrassicin, GBN = glucobrassicanapin, GBC = glucobrassicin, NAS = gluconasturtiin, 4ME = 4-methoxy-glucobrassicin, NEO = neoglucobrassicin, *B.n.* = *Brassica napus*, *B.r.* = *Brassica rapa*, *S.a.* = *Sinapis alba*, *R.s.* = *Raphanus sativa*, *B.o.* = *Brassica oleracea*

Cultivar (species)	PRO[1]	SIN[1]	GNL[1]	RAA[1]	SIB[3]	GNA[1]	4OH[2]	GBN[1]	GBC[2]	NAS[3]	4ME[2]	NEO[2]	**Total**
Robust (*B.n.*)	0.1	0.0	0.0	0.0	0.0	0.4	0.0	0.1	0.1	0.1	0.0	0.0	**0.7**
Robust (*B.n.*)	0.2	0.0	0.0	0.0	0.1	0.0	0.0	0.2	0.2	0.0	0.0	0.0	**0.8**
Robust (*B.n.*)	0.0	0.0	0.0	0.0	0.0	0.0	0.0	0.0	0.3	0.3	0.0	0.0	**0.7**
Campala (*B.n.*)	0.4	0.0	0.0	0.0	0.0	0.4	0.1	1.2	0.4	0.3	0.0	0.0	**2.9**
Grizzly (*B.n.*)	0.0	0.0	0.0	0.0	0.0	0.9	0.2	2.2	0.5	0.8	0.0	0.0	**4.6**
Emerald (*B.n.*)	2.5	0.2	0.0	0.0	0.0	0.9	0.1	1.2	0.6	0.4	0.0	0.0	**6.0**
Lembkes N. (*B.n.*)	0.0	0.0	0.0	0.0	0.0	0.6	0.0	1.5	0.1	0.2	0.0	0.0	**2.5**
Perko (*B.r.*)	0.0	0.0	0.0	0.0	0.0	0.2	0.4	9.2	0.1	1.8	0.0	0.1	**11.8**
Malwira (*B.r.*)	0.3	0.0	0.1	0.0	0.4	0.1	0.0	6.0	0.2	0.2	0.1	0.0	**7.6**
Largo (*B.r.*)	0.1	0.2	0.0	0.0	0.0	0.2	0.3	0.7	0.1	0.4	0.0	0.0	**2.1**
Salut (*B.r.*)	0.0	0.0	0.0	0.0	0.0	0.8	0.3	3.4	0.0	0.5	0.0	0.0	**5.0**
Accent (*S.a.*)	0.3	0.1	0.0	0.0	11.7	0.0	0.0	0.0	0.0	0.0	0.3	0.0	**12.4**
Adagio (*R.s.*)	0.1	0.1	0.0	0.0	0.0	0.1	0.0	0.0	1.9	2.7	0.1	0.0	**4.9**
Brunswijker (*B.o.*)	5.2	0.2	0.1	1.6	0.0	0.0	0.1	0.1	3.0	0.3	0.0	0.0	**10.5**

Overall, based on their GSL the tested cultivars can be subdivided into four classes: (1) in oilseed rape and turnip rape aliphatic compounds predominated. (2) white mustard cv. 'Accent' contained mainly the aromatic compound sinalbin. (3) in oilseed radish cv. 'Adagio' indole and aromatic substances predominated. (4) white cabbage contained mainly aliphatic and indole substances. No correlations between single GSL compounds and the feeding preference of adult cabbage stem flea beetle were detected.

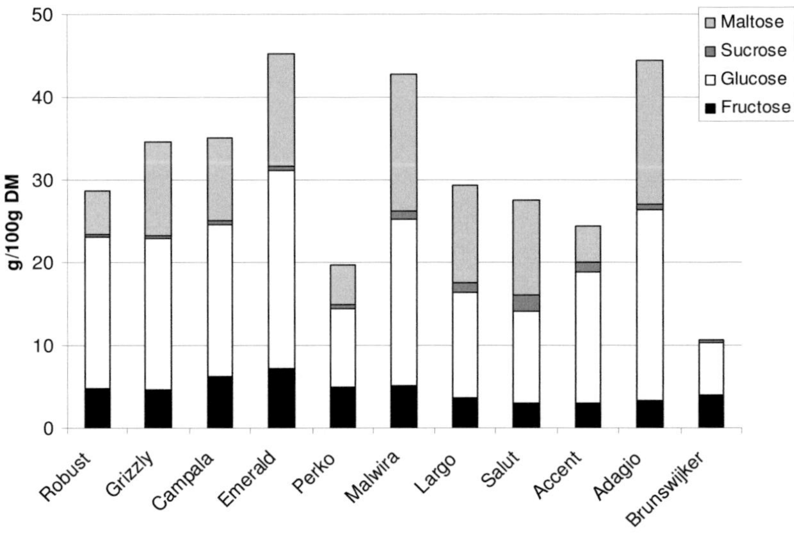

Fig. 2: Content of glucose, fructose, sucrose and maltose (g/100g DM) in leaf lamina of plant species and cultivars used for dual choice tests

The content and composition of sugars varied widely between species and cultivars (Fig. 2). Glucose predominated in the leaf lamina of the tested plants, with concentrations ranging from 6.0 to 22.7g/100g DM in white cabbage and 'Emerald', respectively. Fructose was detected in all tested plants but in contrast to glucose the concentration in 'Salut' and 'Emerald' varied from 2.8 to 6.9g/100g DM, respectively. In all tested plants sucrose was present in very small amounts and with a low variability. Maltose was not detected in white cabbage but was present in all other species and cultivars. The concentration of maltose was highest in oilseed radish with 16.6 g/100g DM. A very high content of maltose was also found in fodder rape 'Emerald' and turnip rape 'Malwira'. In spite of the huge variation between the sugar contents of the tested plants there was no correlation between the sugar content in the leaf lamina and the feeding preference of adult CSFB.

The content of phenolic compounds was only little different between the tested plants. No correlation was found between the concentration of total phenolics and the feeding preference of the beetles throughout our experiments (Fig. 3).

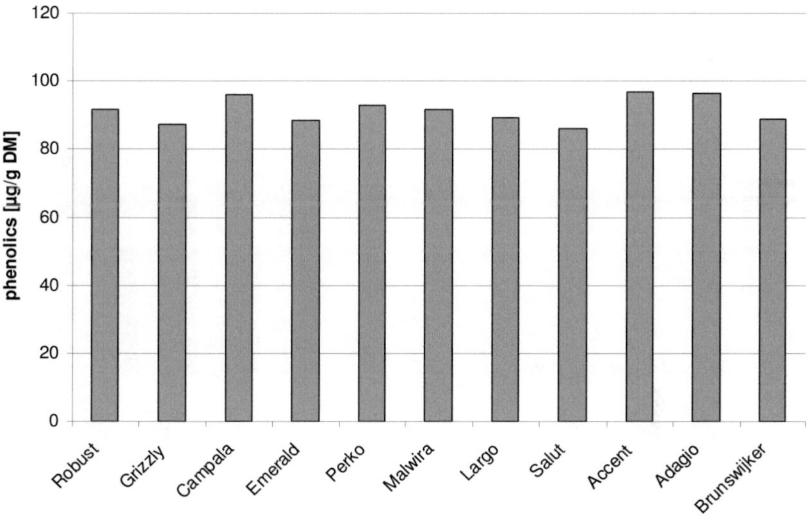

Fig. 3: Mean content of total phenolics (µg/g DM) in leaf lamina of plant species and cultivars used for dual choice tests

The trichome density on the upper leaf surface showed substantial differences between brassicaceous species but these were not statistically significant (Fig. 4). However, the trichome density was positively correlated to the feeding preference of CSFB when all plants were included into the statistical analysis (df=10; r^2=0.44; p=0.0357). Within cultivars of oilseed rape and turnip rape this correlation was not detected.

The dry matter weight of the leaf lamina varied widely between the tested plant species and cultivars (Fig. 4). The highest weight of dry matter was found in the oilseed rape cultivar 'Lembkes Normal' while the lowest dry matter weight occurred in the oilseed rape cultivar 'Campala'. Despite huge differences between the tested cultivars, the dry matter weight was not correlated to the feeding preference of adult CSFB.

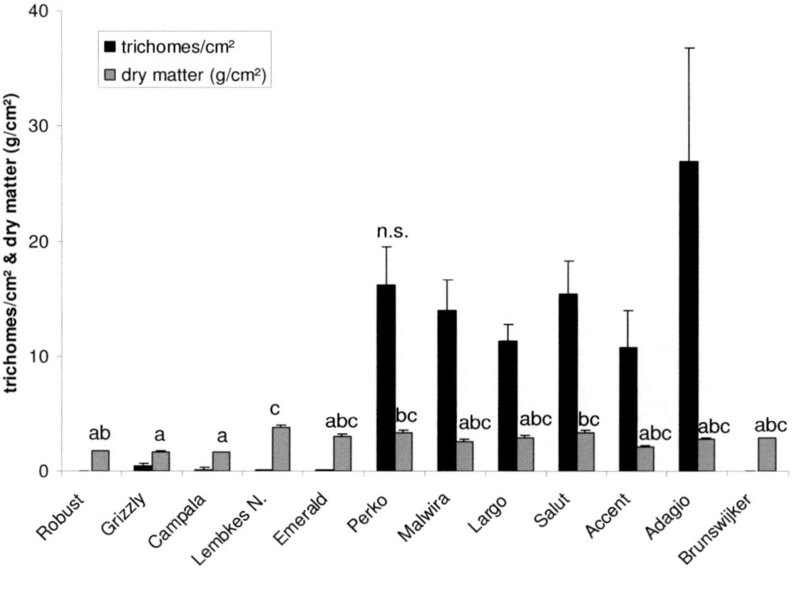

Fig. 4: Mean trichome density (trichomes/cm²) and dry matter weight (g/cm²) of leaf lamina of plant species and cultivars used for dual choice tests; different letters indicate significant differences ($p \leq 0.05$) within dry matter content of leaf laminas

Discussion

The novel design of our dual choice test was succesfully applied to overcome disadvantages of dual or multiple choice tests reported in earlier studies. In the past, the feeding behaviour of adult CSFB was often investigated by using artificial agar media or excised leaves. Results obtained by using these methods are hardly comparable with results of behavioural studies on intact plants. Primary and secondary leaf metabolites are more diverse than in any artificial media

(Schoonhovenet al., 2005). Even though, an artificial medium has potential to identify the effect of specific plant substances as shown by Bartlet et al. (1994). In addition, due to the destruction of plant cells substances such as glucosinolates can be degraded when using excised leaves in feeding experiments (Hopkins et al., 2009).

Adult CSFB significantly preferred turnip rape and oilseed radish to oilseed rape 'Robust' in our experiments. A similar preference of CSFB has been reported in the literature: In field experiments CSFB preferred turnip rape to oilseed rape and also discriminated between different cultivars of oilseed rape (Williams, 1989; Lambdon et al., 1998). In contrast to the feeding behaviour of many other insects specialized on brassicaceous host plants, CSFB did not avoid white mustard cv. 'Accent' when it was offered together with oilseed rape in our experiments. In laboratory tests *Phyllotreta* spp. preferred oilseed rape to white mustard (Bodnaryk and Palaniswamy, 1990; Ekbom, 1995). Similar to our results, the feeding preference of *Mamestra configurata* larvae was inconsistent when white mustard was offered with oilseed rape (Ulmer et al., 2001). Previous studies have shown that similar to CSFB the adult cabbage stem weevil (*Ceutorhynchus pallidactylus* Marsh.) preferred oilseed rape leaves to white cabbage leaves and consumed a higher leaf area on oilseed rape (Eickermann and Ulber, 2010).

Although much work has been done on this topic, plant cues responsible for differences in the feeding preference are not yet fully understood. From the literature it is known that plant substances derived from primary and secondary plant metabolism can influence the behaviour of herbivorous insects (Schoonhoven et al., 2005). Many researchers have focused on the role of GSL, the most important group of secondary plant substances in Brassicaceae (Hicks, 1974; Bartlet et al., 1996; Siemens and MitchellOlds, 1996; Bodnaryk, 1997). There was no clear correlation between the content of single or total GSL and the feeding preference of adult CSFB in our experiments. However, it has been reported in the literature that GSL act as feeding stimuli and are required for host plant acceptance by CSFB (Bartlet and Williams, 1991). The impact of individual GSL compounds or blends of several compounds on the feeding behaviour is not clear. In field experiments CSFB caused significantly higher feeding damage to *Brassica napus* lines containing high GSL content than to lines containing a lower content of GSL (Giamoustaris and Mithen, 1995). In these experiments, gluconapin was found to be more attractive than glucobrassicanapin or progoitrin (Giamoustaris and Mithen, 1995). Under laboratory conditions, the feeding activity of adult CSFB was not correlated to the total GSL content of different oilseed rape lines (Bartlet et al., 1996). When single GSL compounds were added to an artificial agar medium, sinigrin was less preferred by CSFB than glucobrassicin; different compositions of GSL had only little effect on adult feeding (Bartlet et al., 1994). For crucifer flea beetle (*Phyllotreta cruciferae*) individual GSL compounds and their concentration had little effect on feeding by adults (Bodnaryk and Palaniswamy, 1990). However, the leaf consumption was highest at intermediate levels of total GSL (Bodnaryk and Palaniswamy, 1990; Siemens and MitchellOlds, 1996).

Another important group of secondary plant substances, the phenolics, is also known to potentially affect the feeding behaviour of herbivorous insects like *Phyllotreta armoraciae*, *P. nemorum* and *P. cruciferae* on Brassicaceae (Nielsen, 1978; Bernays, 1981; Onyilagha et al., 2004; Schoonhoven et al., 2005) However, some insects like *Spodoptera littoralis* or M*amestra configurata* do not respond to varying levels of phenolic compounds (Hartley, 1988; McCloskey and Isman, 1995; Johnson and Felton, 2001). In our set of test plants the content of phenolic compounds was only little different, as also has been described for other brassicaceous species (Cole, 1994). There is a large diversity of plant phenolics, and single phenolic compounds can vary considerably between species of Brassicaceae, as reported from radish and oilseed rape (Onyilagha et al., 2003). Due to the very homogenous content of total phenolics in our set of test plants, there was no correlation between the feeding preference and the content of phenolics. In our experiments it has not been tested whether single phenolics, such as flavanoids, have a significant impact, as shown for *Mamestra configurata* (Onyilagha et al., 2004). In future studies, a detailed analysis of the effects of single phenolic compounds might reveal factors which are responsible for the feeding preference of CSFB.

Sugars have been reported to stimulate plant feeding of different insect species (Schoonhoven et al., 2005). The sugar content and the performance of herbivorous insect *Delia radicum* on the roots of *Brassica* species were positively correlated (Hopkins et al., 1999; Schoonhoven et al., 2005). Sucrose, fructose and glucose were reported to be the most effective feeding stimulants for CSFB and other insect species (Bartlet et al., 1994; Schoonhoven et al., 2005). However, these experiments were conducted by using an artificial agar medium with pure sugars added and the effect of sugar may be less distinct in plants. In our study, the lack of correlations between the content of sugar and the feeding preference might be due to the low variation of sugars in the tested plant species and cultivars which was too small to impact feeding of CSFB. The effect of sugars on the feeding behaviour of insects increases with increasing concentrations (Schoonhoven et al., 2005). Therefore the reduction of sugar content in the leaves might decrease the feeding activity of insect herbivores.

In our experiments other plant cues such as morphological structures might have influenced the feeding behaviour, as well (Schoonhoven et al., 2005). Morphological structures on the plant surface such as trichomes or epicuticular waxes can play an important role for host plant acceptance by insects (Schoonhoven et al., 2005). Trichomes may act as mechanical barrier for small insects. By lowering host plant acceptance the pubescence of leaves can provide a mechanism of resistance against insects,, as has been reported for *Pieris rapae* feeding on *Brassica rapa* (Agren and Schemske, 1993; Schoonhoven et al., 2005). The trichome density of individual plant species is determined genetically and may be affected by external factors such as insect herbivory (Baur et al., 1991; Agren and Schemske, 1994; Traw and Dawson, 2002). Recent work confirms our findings that the trichome density differs substantially between oilseed rape, cabbage and white mustard (Le Guigo et al., 2011). We observed a positive correlation between the feeding preference of CSFB and trichome density only when all tested plants were included in the statistical analysis. There was no correlation between feeding of CSFB and trichome density within cultivars of oilseed rape and turnip rape. Therefore we suppose that other plant characteristics are more important for the feeding preference of CSFB on these species. Earlier studies dealing with the effect of trichome density on feeding behaviour of crucifer flea beetle showed that *P. cruciferae* preferred hairless host plants to pubescent plants. Hairy surfaces even provide a resistance factor to feeding of flea beetles (Lamb, 1980; Soroka et al., 2011). As crucifer flea beetles are much smaller than cabbage stem flea beetles they are probably affected more severely by leaf trichomes. Further, the type of trichomes might have played a role for feeding of CSFB in our experimens. Lamb (1980) found that soft trichomes on turnip rape do not deter cabbage flea beetle feeding as much as stiff hairs do.

We further suppose that CSFB consume a larger area from leaves of lower dry weight for ingesting the same amount of food as provided by leaves of a higher dry weight. In contrast to earlier studies on canola (Ulmer et al., 2001), we found significant differences between the dry weight of the tested oilseed rape cultivars 'Lembkes Normal', 'Robust', 'Grizzly' and 'Campala'. However, the feeding preference of CSFB was not significantly correlated with the dry weight of leaves.

Synergistic effects between chemical plant factors may play a significant role for the feeding preference, as has been described for sucrose and sinigrin (Bartlet et al., 1994). Furthermore, the concentration of plant nutrients, such as zinc, phosphorus or potassium within the plant tissue was found to impact food consumption by *Otiorhynchus sulcatus* (Coyle et al., 2010). In addition, the wax layer on the leaf surface might protect the plants against herbivores and needs to be considered (Stadler and Reifenrath, 2009). In field experiments the amount of leaf wax of *Brassica* species was negatively correlated to CSFB and crucifer flea beetle feeding damage (Bodnaryk, 1992; Lambdon et al., 1998).

Our results show that white cabbage might have potential to be used for breeding of oilseed rape cultivars which are less susceptible to insect feeding. However, the factors which determine the feeding preference of CSFB need to be investigated in greater detail.

References

Adams JB, van Emden HF, 1972. The biological properties of aphids and their host plant relationships. In: Aphid technology. (ed. H F van Emden). Academic Press Inc. Ltd., London.

Agren J, Schemske DW 1993. The cost of defense against herbivores - an experimental-study of trichome production in *Brassica rapa*. American Naturalist 141, 338-350.

Agren J, Schemske DW 1994. Evolution of trichome number in a naturalized population of *Brassica rapa*. American Naturalist 143, 1-13.

Alford DV, Nilsson C, Ulber B, 2003. Insect pests of oilseed rape crops. In: Biocontrol of oilseed rape pests. (ed. by D V Alford). Blackwell Science Ltd., Oxford, 9-42.

Barari H, Cook SM, Clark SJ, Williams IH 2005. Effect of a turnip rape (*Brassica rapa*) trap crop on stem-mining pests and their parasitoids in winter oilseed rape (*Brassica napus*). Biocontrol 50, 69-86.

Bartlet E, Mithen R, Clark SJ 1996. Feeding of the cabbage stem flea beetle *Psylliodes chrysocephala* on high and low glucosinolate cultivars of oilseed rape. Entomologia Experimentalis et Applicata 80, 87-89.

Bartlet E, Parsons D, Williams IH, Clark SJ 1994. The influence of glucosinolates and sugars on feeding by the cababge stem flea beetle, *Psylliodes chrysocephala*. Entomologia Experimentalis et Applicata 73, 77-83.

Bartlet E, Williams IH 1991. Factors restricting the feeding of the cabbage stem flea beetle (*Psylliodes chrysocephala*). Entomologia Experimentalis et Applicata 60, 233-238.

Baur R, Binder S, Benz G 1991. Nonglandular leaf trichomes as short-term inducible defense of the gray alder, *Alnus incana* L., against the chrysomelid beetle, *Agelastica alni* L. Oecologia 87, 219-226.

Bernays EA 1981. Plant tannins and insect herbivores - an appraisal. Ecological Entomology 6, 353-360.

Birch ANE 1988. Field and glasshouse studies on the components of resistance to root fly attack in swedes. Annals of Applied Biology 113, 89-100.

Bodnaryk RP 1992. Leaf epicuticular wax, an antixenotic factor in brassicaceae that affects the rate and pattern of feeding of flea beetles, P*hyllotreta cruciferae* (goeze). Canadian Journal of Plant Science 72, 1295-1303.

Bodnaryk RP 1997. Will low-glucosinolate cultivars of the mustards *Brassica juncea* and *Sinapis alba* be vulnerable to insect pests? Canadian Journal of Plant Science 77, 283-287.

Bodnaryk RP, Palaniswamy P 1990. Glucosinolate levels in cotyledons of mustard, B*rassica juncea* L. and rape, *Brassica napus* L. do not determine feeding of flea beetle *Phyllotreta cruciferae* (goeze). Journal of Chemical Ecology 16, 2735-2746.

Bommarco R, Miranda F, Bylund H, Bjorkman C 2011. Insecticides suppress natural enemies and increase pest damage in cabbage. Journal of Economic Entomology 104, 782-791.

Büchi R 1995. Combination of trap plants (*Brassica rapa var. Vestris*) and insecticide use to control rape pests. IOBC/wprs Bulletin 18 (4), 102-119.

Carcamo H, Olfert O, Dosdall L, Herle C, Beres B, Soroka J 2007. Resistance to cabbage seedpod weevil among selected Brassicaceae germplasm. Canadian Entomologist 139, 658-669.

Cleemput S, Becker HC 2011. Genetic variation in leaf and stem glucosinolates in resynthesized lines of winter rapeseed (B*rassica napus* L.). Genetic Resources and Crop Evolution. doi:10.1007/s10722-011-9701-x.

Cole RA 1994. Locating a resistance mechanism to the cabbage aphid in two wild brassicas. Entomologia Experimentalis et Applicata 71, 23-31.

Coyle DR, Clark KE, Raffa KF, Johnson SN 2010. Prior host feeding experience influences ovipositional but not feeding preference in a polyphagous insect herbivore. Entomologia Experimentalis et Applicata 138, 137-145.

Eickermann M, Ulber B 2010. Screening of oilseed rape and other brassicaceous genotypes for susceptibility to *Ceutorhynchus pallidactylus* (Mrsh.). Journal of Applied Entomology 134, 542-550.

Eickermann M, Ulber B, Vidal S 2010. Resynthesized lines and cultivars of *Brassica napus* L. provide sources of resistance to the cabbage stem weevil (*Ceutorhynchus pallidactylus* (Marsh.)). Bulletin of Entomological Research 110, 1-8.

Ekbom B 1995. Flea beetle (*Phyllotreta undulata*) feeding preference. IOBC/wprs Bulletin 18 (4), 88-94.

Evans KA, Scarisbrick DH 1994. Integrated insect pest-management in oilseed rape crops in Europe. Crop Protection 13, 403-412.

Giamoustaris A, Mithen R 1995. The effect of modifying glucosinolate content of leaves of oilseed rape (B*rassica napus* ssp. o*leifera*) on its interaction with specialist and generalist pests. Annals of Applied Biology 126, 347-363.

Hartley SE 1988. The inhibition of phenolic biosynthesis in damaged and undamaged birch foliage and its effect on insect herbivores. Oecologia 76, 65-70.

Heimbach U, Müller A, 2011. Pyrethroid resistance of oilseed rape pest insects from 2005-2010 and german insecticide resistance management strategy. In: 13th International Rapeseed Congress Prague, 1278-1281.

Hicks KL 1974. Mustard oil glucosides - feeding stimulants for adult cabbage flea beetles, *Phyllotreta cruciferae* (Coleopteran; Chrysomelidae). Annals of the Entomological Society of America 67, 261-264.

Hopkins RJ, Griffiths DW, McKinlay RG, Birch ANE 1999. The relationship between cabbage root fly (*Delia radicum*) larval feeding and the freeze-dried matter and sugar content of brassica roots. Entomologia Experimentalis et Applicata 92, 109-117.

Hopkins RJ, van Dam NM, van Loon JJA 2009. Role of glucosinolates in insect-plant relationships and multitrophic interactions. Annual Review of Entomology 54, 57-83.

Hoxie RP, Wellso SG, Webster JA 1975. Cereal leaf beetle response to wheat trichome length and density. Environmental Entomology 4, 365-370.

Johnson KS, Felton GW 2001. Plant phenolics as dietary antioxidants for herbivorous insects: a test with genetically modified tobacco. Journal of Chemical Ecology 27, 2579-2597.

Keutgen AJ, Pawelzik E 2008. Influence of pre-harvest ozone exposure on quality of strawberry fruit under simulated retail conditions. Postharvest Biology and Technology 49, 10-18.

Lamb RJ 1980. Hairs protect pods of mustard (*Brassica hirta* Gisilba) from flea beetle feeding damage. Canadian Journal of Plant Science 60, 1439-1440.

Lambdon PW, Hassall M, Mithen R 1998. Feeding preferences of woodpigeons and flea-beetles for oilseed rape and turnip rape. Annals of Applied Biology 133, 313-328.

Larsen L, J. Nielsen, A. Plöger & H. Sorensen, 1985. Response of some beetle species to varieties of oilseed rape and to pure glucosinolates. In: World crops: Production, utilization, description. (ed. by H Sorensen). Martinus Nijhoff Publishers, Brussels, 230-244.

Le Guigo P, Qu Y, Le Corff J 2011. Plant-mediated effects on a toxin-sequestering aphid and its endoparasitoid. Basic and Applied Ecology 12, 72-79.

McCloskey C, Isman MB 1995. Plant-growth stage effects on the feeding and growth-responses of the bertha armyworm, *Mamestra configurata* to canola and mustard foliage. Entomologia Experimentalis et Applicata 74, 55-61.

Nielsen JK 1978. Host plant-selection of monophagous and oligophagous flea beetles feeding on crucifers. Entomologia Experimentalis et Applicata 24, 562-569.

Nilsson C 1990. Yield losses in winter oilseed rape caused by cabbage stem flea beetle larvae (*Psylliodes chrysocephala* (L.)). IOBC/wprs Bulletin 13 (4), 53-55.

Onyilagha J, Bala A, Hallett R, Gruber M, Soroka J, Westcott N 2003. Leaf flavonoids of the cruciferous species, *Camelina sativa*, *Crambe* spp., *Thlaspi arvense* and several other genera of the family Brassicaceae. Biochemical Systematics and Ecology 31, 1309-1322.

Onyilagha JC, Lazorko J, Gruber MY, Soroka JJ, Erlandson MA 2004. Effect of flavonoids on feeding preference and development of the crucifer pest *Mamestra configurata* Walker. Journal of Chemical Ecology 30, 109-124.

Schoonhoven LM, van Loon JJA, Dicke M, 2005. Insect-plant biology. Oxford University Press, Oxford

Siemens DH, MitchellOlds T 1996. Glucosinolates and herbivory by specialists (Coleoptera: Chrysomelidae, Lepidoptera: Plutellidae): consequences of concentration and induced resistance. Environmental Entomology 25, 1344-1353.

Singleton VL, Rossi JA 1965. Colorimetry of total phenolics with phosphomolybdic-phosphotungstic acid reagents. American Journal of Enology and Viticulture 16, 144-158.

Soroka JJ, Holowachuk JM, Gruber MY, Grenkow LF 2011. Feeding by flea beetles (Coleoptera: Chrysomelidae; *Phyllotreta* spp.) is decreased on canola *(Brassica napus*) seedlings with increased trichome density. Journal of Economic Entomology 104, 125-136.

Stadler E, Reifenrath K 2009. Glucosinolates on the leaf surface perceived by insect herbivores: Review of ambiguous results and new investigations. Phytochemistry Reviews 8, 207-225.

Tansey JA, Dosdall LM, Keddie A, Fletcher RS, Kott LS 2010. Antixenosis and antibiosis resistance to *Ceutorhynchus obstrictus* in novel germplasm derived from *Sinapis alba* x *Brassica napus*. Canadian Entomologist 142, 212-221.

Traw MB, Dawson TE 2002. Differential induction of trichomes by three herbivores of black mustard. Oecologia 131, 526-532.

Ulmer B, Gillott C, Erlandson M 2001. Feeding preferences, growth, and development of *Mamestra configurata* (Lepidoptera: Noctuidae) on Brassicaceae. Canadian Entomologist 133, 509-519.

Williams IH, 2004. Advances in insect pest management of oilseed rape in Europe. In: Insect pest management. (ed. by A R Horowitz, I Ishaaya). Springer-Verlag, Berlin Heidelberg, 181-208.

Williams IH, 2010. The major insect pests of oilseed rape in Europe and their management: an overview. In: Biocontrol-based integrated management of oilseed rape pests. (ed. by I H Williams). Springer Science+Business B.V., Dordrecht, Heidelberg, London, New York, 1-43.

Williams IH 1989. Pest incidence on single low and double low oilseed rape. Aspects of Applied Biology 23, 277-286.

Williams IH, Bartlet E 1993. Host plant affecting the feeding of cabbage stem flea beetle (*Psylliodes chrysocephala* L.) on oilseed rape. IOBC/wprs Bulletin 16 (9), 202-209

Chapter IV: Susceptibility of turnip rape cultivars to infestation by insect pests

Abstract

Turnip rape (*Brassica rapa*) has been reported to be a suitable trap crop for management of specialist insect pests on oilseed rape (*Brassica napus*). Due to the higher attractiveness for many pest species compared to oilseed rape integration of turnip rape has been recommended for reducing pest invasion into crops of oilseed rape. The aim of our investigation was to identify cultivars of turnip rape which are highly attractive for major oilseed rape pests and thus may be used in a trap cropping system. In addition to turnip rape grown in pure stand, seed mixtures of oilseed rape and turnip rape were assessed for their potential to reduce pest infestation on oilseed rape plants. In replicated field plot experiments, the level of infestation of four turnip rape cultivars 'Largo', 'Prisma', 'Malwira' and 'Perko' and two mixed cropping treatments (oilseed rape : turnip rape 95:5 and 90:10) by major insect pests was compared to infestation levels of oilseed rape pure stand (cv. 'Oase' or 'Robust'). Numbers of cabbage stem flea beetle larvae (*Psylliodes chrysocephala*) were significantly higher on turnip rape than on oilseed rape. The numbers of larvae did not differ between the tested cultivars of turnip rape. Pollen beetles (*Meligethes aeneus*) preferred turnip rape to oilseed rape during the early immigration phase. This preference became less obvious at advanced growth stages of both plant species. Significantly higher numbers of adult pollen beetles were found on turnip rape cultivar 'Malwira' than on 'Largo', 'Prisma' and 'Perko'. Infestation by rape stem weevil (*Ceutorhynchus napi*) and cabbage stem weevil (*Ceutorhynchus pallidactylus*) did not differ between the tested *Brassica* species, while larval infestation of cabbage seed weevil (*Ceutorhynchus obstrictus*) was significantly higher in pods of oilseed rape than in pods of turnip rape. Mixed cropping of oilseed rape with turnip rape did not significantly reduce the infestation level and the damage caused by insect pests as compared to oilseed rape in pure stands. Therefore, mixed cropping did not represent a suitable system for integrated management of insect pests on oilseed rape. Growing of the turnip rape cultivar 'Malwira' in perimeter strips can be an option to increase the efficacy of turnip rape as trap crop.

Introduction

Six major insect pests regularly cause significant economic yield losses in crops of winter oilseed rape in Europe (Williams, 2010). They attack the crop in a sequence during the whole growing season (Winfield, 1992). In August and September, the emerging plants are exposed to feeding of adult cabbage stem flea beetle (*Psylliodes chrysocephala*). Later in the season larvae feed within petioles and stems (Kaufmann, 1941). In spring, rape stem weevil (*Ceutorhynchus napi*) and cabbage stem weevil (*Ceutorhynchus pallidactylus*) colonizes the oilseed rape crop for maturation feeding and oviposition (Günthart, 1949). Females of rape stem weevil and cabbage stem weevil deposit their eggs mainly into stems or into stems and petioles, respectively (Williams, 2010). When temperature exceeds 12-15°C, the pollen beetle (*Meligethes aeneus*) colonizes the oilseed rape crops to feed on pollen (Fritzsche, 1957), thereby causing damage to buds and bud abortion (Williams & Free, 1978). At the end of the bud stage and early flowering, cabbage seed weevil and brassica pod midge colonize the crop (Williams, 2010). Severe yield losses are caused by the larvae of both species which develop inside the pods (Alford et al., 2003). Current control of these pests strongly relies on synthetic insecticides (Winfield, 1992; Williams, 2004). Non-chemical pest management strategies have been evaluated during recent years. Conservation of natural enemies, e.g. by reduced tillage systems, seems to provide a promising strategy (Nilsson, 2010). Trap cropping is another option for non-chemical pest management, which has been successfully applied in various crops (Hokkanen, 1991; Shelton & Badenes-Perez, 2006). Trap crops are highly attractive plant species that are able to lure the pests away from the main crop (Hokkanen, 1991), thereby reducing pest infestation of the main crop and pest damage (Hokkanen, 1991). In studies conducted in Switzerland and the UK, turnip rape was preferred to oilseed rape by many specialist insect pests (Büchi, 1990; Lambdon et al., 1998; Barari et al., 2005; Cook et al., 2006). The higher attractiveness of turnip rape for pollen beetle mainly resulted from its earlier development in spring and from olfactory stimuli that are more attractive to pests (Evans and Allen-Williams,

1994; Frearson et al., 2005; Cook et al., 2007). In field trials turnip rape was applied to divert pests during the vulnerable growth stage of oilseed rape (Hokkanen, 1991; Cook, 2002). In many experiments, turnip rape was sown as a perimeter strip at the edges of oilseed rape fields (Cook et al., 2004; Barari et al., 2005; Carcamo et al., 2007; Büchs, 2009). Mixed cropping of oilseed rape with turnip rape was assessed for its potential as pest management strategy as well, often showing variable efficacies of turnip rape as trap crop (Büchi, 1990; Nilsson, 2004). The efficacy of the trap cropping strategy needs to be improved by targeted selection of most suitable turnip rape cultivars (Williams, 2004). An ideal trap crop for pollen beetle management should have flowers and buds at the same time, to provide optimal resources for feeding and oviposition (Frearson et al., 2005). An optimal trap crop was defined as a plant which will be accepted as host plant at the first landing of an insect (Finch and Collier, 2000). In our study, in three year field experiments three turnip rape cultivars were screened for their attractiveness to major insect pests in oilseed rape and their suitability as trap crops. Furthermore, intercropping of oilseed rape and turnip rape was assessed for its potential as an integrated pest management strategy.

Material and Methods

Experimental setup of field trials

Field experiments were conducted in three growing seasons from August 2008 until July 2011 at two locations in central Germany: Lower Saxony near Goettingen and Hesse near Witzenhausen. Test cultivars of turnip rape and oilseed rape were selected according to their differential phenology, especially during the flowering period. In the first year the oilseed rape cultivar 'Oase' and the turnip rape cultivars 'Largo', 'Prisma' and 'Perko PVH' were grown. Based on these results, the oilseed rape cultivar 'Robust' and the turnip rape cultivars 'Largo', 'Malwira', and 'Perko PVH' were used in the following two experimental seasons. All cultivars had a very low content of erucic acid and glucosinolates in their seeds except 'Perko PVH' and 'Malwira' which had high contents of both substances. Depending on the weather conditions and soil moisture, plots were sown between mid of August and beginning of September. All cultivars were sown in pure stands. Additionally, the oilseed rape cultivars 'Oase' (2008) and 'Robust' (2009 & 2010) were sown with turnip rape 'Largo' in seed mixtures of 95:5 and 90:10 (oilseed rape : turnip rape). Sowing rate was 53 seeds per m² in all treatments. Plots (15m long, 7.5m wide) were arranged in a complete randomized block design with four replications of each cultivar and mixture. To reduce edge effects, experiments were bordered by a 3m wide perimeter strip of oilseed rape of the same cultivar as used in the experiment. Crop management followed the guidelines for organic farming, without application of pesticides. Weeds were controlled twice a year by hoeing. Immigration of insect pests into the experimental area was monitored by yellow water traps in autumn and in spring (results in Appendix).

Data collection

The level of pest infestation was assessed on plants of pure stand plots and on oilseed rape plants of seed mixture plots. The number of larvae of cabbage stem flea beetle, rape stem weevil and cabbage stem weevil was assessed in 25 plants randomly collected from each plot. For assessment of cabbage stem flea beetle and stem weevils, plants were collected at growth stage BBCH 29 and BBCH 69 (Lancashire et al., 1991), respectively, before larvae started to migrate to soil for pupation. The plants were dissected in the laboratory and larvae were identified to species level.

At growth stage BBCH 69 the plant height and length of feeding galleries were measured to estimate the damage caused by stem mining larvae. The level of pollen beetle infestation was assessed at regular intervals from the beginning of bud stage (BBCH 51) to main flowering (BBCH 65). Adult beetles were beaten from the inflorescences into a plastic tray (26.5cm * 33.5cm) and counted. In each plot, 50 plants were chosen randomly. Before ripening (BBCH 69 to 81), the main raceme and 2nd lateral racemes of 15 to 25 plants were collected randomly. The number of pods and of empty stalks was assessed to estimate percent bud loss caused by pollen beetles. At BBCH 81, the number of pods showing exit holes caused by seed weevil larvae was counted on the main and 2nd lateral raceme of

15 to 25 plants. The seed yield of oilseed rape in pure stand and of seed mixtures was measured by using a parcel combine.

Data analysis

The stem injury index was calculated as an indicator for the damage caused by stem mining weevils. The length of feeding galleries was divided by stem length (Eickermann et al., 2010). The bud loss index was calculated by dividing the number of empty stalks, caused by pollen beetle feeding, by the total number of empty stalks and pods. In each year, data on the infestation level of cabbage stem flea beetle, rape stem weevil, cabbage stem weevil and cabbage seed weevil as well as the stem injury index, bud loss index and seed yield of both experimental locations were combined for statistical analyses. ANCOVA was conducted with site as covariant, cultivars or mixture as independent variable and infestation level or damage index as dependent variable. ANOVA was calculated when only data of one experimental site were available. For pollen beetle infestation ANOVA was applied for each single assessment date and each site, with cultivar or mixture as independent variable and beetle numbers per main raceme as dependent variable. ANCOVA as well as ANOVA were followed by post-hoc Tukey HSD test. Differences with $\alpha \leq 0.05$ were regarded as statistically significant. Data of stem injury index and bud loss index were arcsin transformed before statistical analysis. Data with inhomogenous variances were log (x+1) transformed. For all statistical analyses the software package Statistica 9.1 (StatSoft Inc.) was used.

Results

At both experimental sites, the phenology of the tested cultivars showed substantial differences between the beginning of bud stage to flowering (Fig. 1 & 2), the most susceptible stage for pollen beetle damage. The difference between the phenology of oilseed rape and turnip rape was smaller in 2009 when the oilseed rape cultivar ‘Oase’ was used in the experiments. In 2010 and 2011 the difference between the phenology of the oilseed rape cultivar ‘Robust’ and turnip rape was more obvious at both locations. In both years flowering of the cultivar ‘Robust’ started at least 9 days later than flowering of the earliest turnip rape cultivar. No significant difference occurred between the phenology of the turnip rape cultivars 'Largo' and 'Prisma', while the cultivar 'Perko PVH' started flowering first in most experiments. In 2010 and 2011, the turnip rape cultivar 'Malwira' started flowering up to two days later than the other turnip rape cultivars.

During the early bud stage, the oilseed rape cultivar was significantly less infested by pollen beetle than the turnip rape cultivars in most experiments (Tab. 1). When the first turnip rape cultivars started to flower this trend turned to the opposite. Oilseed rape was commonly preferred by pollen beetle when it was still in bud stage whilst the turnip rape was already in flower. Among the turnip rape cultivars, 'Perko PVH' was most attractive in 2009 while pollen beetle numbers were not significantly different between 'Largo' and 'Prisma'. In 2010 and 2011, pollen beetle numbers on the set of turnip rape cultivars showed a higher variation than in 2009. 'Malwira' was most attractive for pollen beetles especially at yellow bud stage (BBCH 59) when the other turnip rape cultivars already were in flower. Similar to the results of 2009, the cultivar 'Perko PVH' tended to attract more pollen beetles per plant than plants of cultivar 'Largo'.

In 2009 and 2010, the infestation by cabbage stem flea beetle differed significantly between the tested Brassica species (Tab. 2). Turnip rape showed significantly higher numbers of larvae than oilseed rape while no significant difference could be detected among larval infestation of the turnip rape cultivars in both years. The infestation level by rape stem weevil larvae and cabbage stem weevil larvae was low in 2009 and increased in 2010 and 2011 (Tab. 2). Despite high levels of infestation in 2010 and 2011, the number of larvae did not differ significantly between oilseed rape and turnip rape or between the tested cultivars of turnip rape. Contrasting to stem weevils, cabbage seed weevil infestation was significantly higher on oilseed rape than on turnip rape in 2010 and 2011 (Tab. 2). Pod infestation by cabbage seed weevil was not significantly different between the turnip rape cultivars. Mixed cropping of oilseed rape and turnip rape had no significant effect on infestation of oilseed rape by cabbage stem flea beetle, rape stem weevil, cabbage stem weevil, pollen beetle and cabbage seed weevil as compared to oilseed rape grown in pure stand (Tab. 3 & 4).

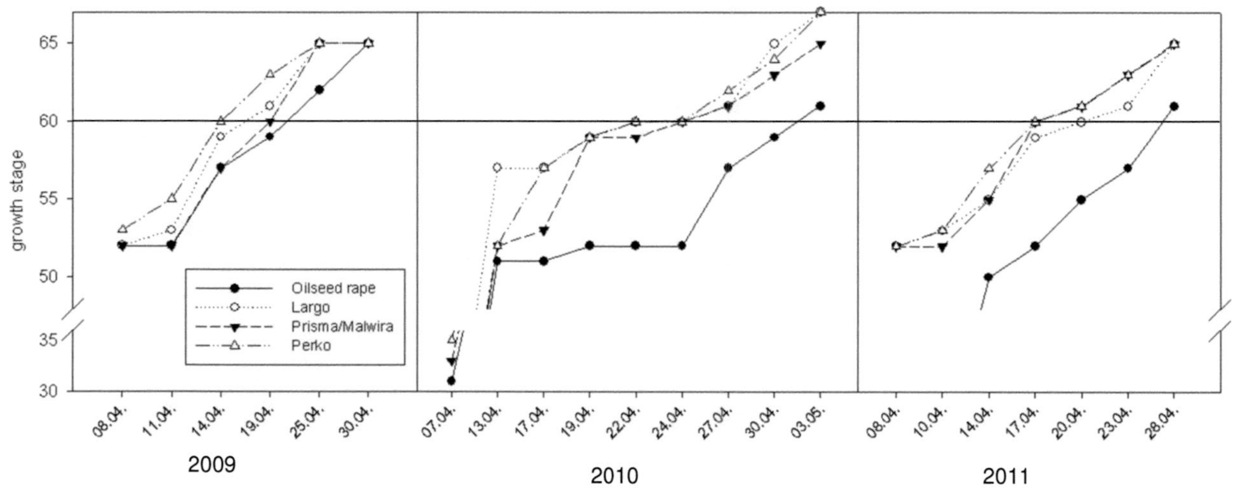

Fig. 1: Plant growth stages (Lancashire et al., 1991) of oilseed rape ('Oase': 2009; 'Robust': 2010, 2011) and turnip rape cultivars 'Largo', 'Prisma' (2009), 'Malwira' (2010 & 2011) & 'Perko PVH' during pollen beetle immigration at Witzenhausen

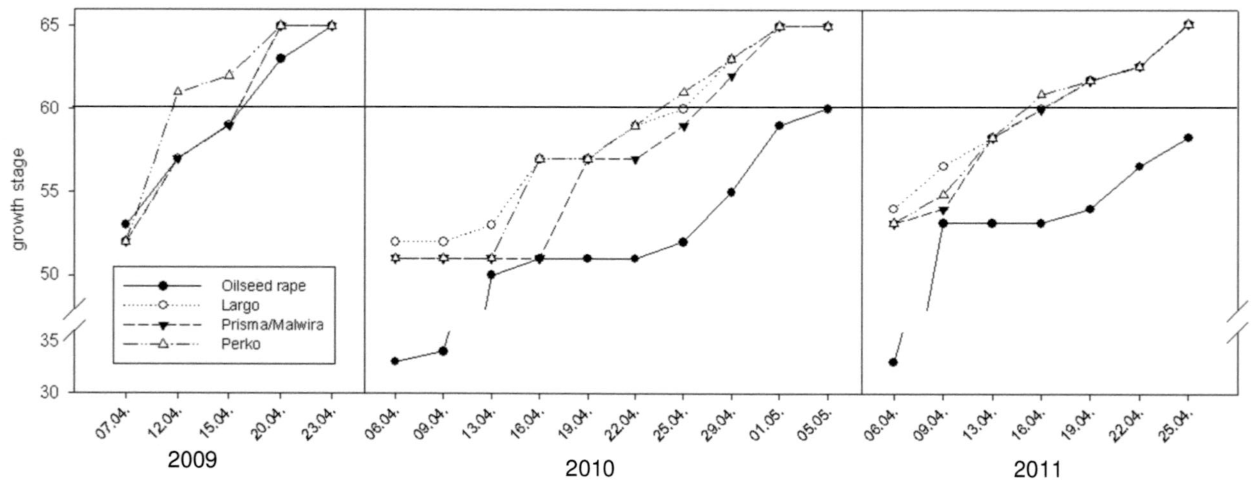

Fig. 2: Plant growth stages (Lancashire et al., 1991) of oilseed rape ('Oase': 2009; 'Robust': 2010, 2011) and turnip rape cultivars 'Largo', 'Prisma' (2009), 'Malwira' (2010 & 2011) & 'Perko PVH' during pollen beetle immigration at Göttingen

Tab. 1: Mean (±SE) infestation of oilseed rape (OSR) ('Oase': 2009; 'Robust': 2010, 2011) and turnip rape cultivars 'Largo', 'Prisma' (Pri) (in 2009),'Malwira' (Ma) (in 2010 & 2011) and 'Perko PVH' by *Meligethes aeneus* during bud stage. Different letters in a row indicate significant differences (α≤0.05)

Year	Goettingen					Witzenhausen				
	Date	OSR	Largo	Pri/Ma	Perko	Date	OSR	Largo	Pri/Ma	Perko
	07.04.	0.8 (±0.2) a	1.5 (±0.5) a	1.3 (±0.2) a	3.6 (±0.6) b	**08.04.**	0.7 (±0.1) a	1.2 (±0.4) a	0.8 (±0.2) a	2.5 (±0.2) b
	09.04.	1.5 (±0.1) a	2.9 (±0.2) b	2.8 (±0.1) b	5.8 (±0.1) c	**11.04.**	2.2 (±0.0) a	2.6 (±0.4) a	2.3 (±0.2) a	6.2 (±0.2) b
	12.04.	3.2 (±0.1) a	4.9 (±0.4) b	4.5 (±0.4) ab	9.5 (±0.3) c	**14.04.**	2.5 (±0.4) a	4.2 (±0.3) a	4.4 (±0.4) a	7.7 (±0.9) b
2009	**15.04.**	4.0 (±0.3) a	5.6 (±0.1) ab	6.0 (±0.7) b	7.1 (±0.6) b	**19.04.**	1.7 (±0.2) a	4.0 (±0.5) a	3.5 (±0.8) a	3.8 (±0.6) a
	20.04.	5.1 (±0.6) ab	5.1 (±0.3) ab	6.6 (±0.4) a	3.8 (±0.4) b	**22.04.**	1.8 (±0.1) a	3.0 (±0.3) b	5.2 (±0.6) c	0.8 (±0.1) d
	23.04.	0.8 (±0.1) a	2.0 (±0.2) b	2.3 (±0.2) b	0.6 (±0.1) a	**25.04.**	0.8 (±0.1) a	1.2 (±0.2) a	1.8 (±0.3) ab	0.6 (±0.1) a
						30.04.	0.4 (±0.1) a	0.1 (±0.0) b	0.3 (±0.1) b	0.1 (±0.0) b
	09.04.	0.2 (±0.0) a	1.2 (±0.1) b	1.8 (±0.2) bc	2.1 (±0.2) c	**07.04.**	0.6 (±0.1) a	2.6 (±0.1) b	2.9 (±0.3) b	2.4 (±0.4) b
	13.04.	0.2 (±0.1) a	1.1 (±0.1) b	2.1 (±0.5) b	2.5 (±0.4) b	**13.04.**	0.4 (±0.0) a	1.7 (±0.1) b	2.0 (±0.3) b	2.5 (±0.3) b
	16.04	0.3 (±0.0) a	1.1 (±0.2) b	1.8 (±0.5) bc	2.5 (±0.3) c	**17.04.**	1.3 (±0.2) a	6.2 (±0.7) b	7.2 (±0.4) bc	8.9 (±0.6) c
	19.04.	0.7 (±0.2) a	7.9 (±0.9) b	10.6 (±1.1) b	11.9 (±0.9) b	**19.04.**	1.4 (±0.1) a	7.7 (±0.3) b	9.6 (±0.7) c	12.1 (±0.3) d
2010	**22.04.**	0.7 (±0.1) a	5.1 (±0.5) b	7.3 (±0.6) b	7.3 (±1.1) b	**22.04.**	1.1 (±0.1) a	4.7 (±0.3) b	6.1 (±0.8) b	6.9 (±0.8) b
	25.04.	5.2 (±0.7) a	14.8 (±2.9) b	25.9 (±4.2) c	16.7 (±0.9) bc	**24.04.**	2.1 (±0.0) a	7.1 (±0.4) b	11.3 (±0.3) c	12.0 (±0.6) c
	29.04.	4.8 (±0.3) a	6.7 (±1.1) a	15.4 (±2.7) b	7.9 (±0.5) a	**27.04.**	5.2 (±0.3) a	7.5 (±0.8) a	14.5 (±0.6) b	10.3 (±0.9) c
	01.05.	6.3 (±0.5) a	3.2 (±0.5) b	6.6 (±0.3) a	3.7 (±0.3) b	**30.04.**	7.2 (±0.8) a	4.0 (±0.3) b	16.1(±1.2) c	5.2 (±0.2) ab
	05.05.	3.8 (±0.4) a	0.5 (±0.1) b	1.1 (±0.2) b	0.7 (±0.2) b	**03.05.**	7.1 (±0.7) a	1.8 (±0.3) b	8.6 (±1.2) a	3.3 (±0.8) b
						08.05.	4.5 (±0.5) a	0.7 (±0.1) b	4.1 (±0.4) a	0.9 (±0.2) b
	06.04.		4.8 (±0.6) a	4.7 (±0.4) a	6.0 (±0.5) a	**08.04.**	0.6 (±0.1) a	9.5 (±0.9) b	8.8 (±0.2) b	9.5 (±0.1) b
	09.04.	0.3 (±0.1) a	5.0 (±0.6) b	5.2 (±0.3) b	5.4 (±0.2) b	**10.04.**	1.4 (±0.2) a	9.1 (±1.0) b	10.8 (±1.5) b	11.4 (±1.6) b
	13.04.	0.4 (±0.1) a	12.3 (±2.6) b	15.0 (±1.2) b	14.3 (±2.6) b	**14.04.**	1.6 (±0.2) a	8.8 (±0.5) b	9.3 (±0.2) bc	10.3 (±0.1) c
2011	**16.04.**	1.3 (±0.4) a	13.8 (±2.4) b	19.2 (±1.0) b	18.3 (±2.0) b	**17.04.**	1.9 (±0.1) a	8.8 (±0.3) b	11.0 (±0.5) c	11.3 (±0.4) c
	19.04.	2.0 (±0.5) a	11.8 (±0.7) b	20.5 (±1.6) c	13.5 (±1.7) b	**20.04.**	5.4 (±0.3) a	10.3 (±0.4) b	13.7 (±0.3) c	10.4 (±0.5) b
	22.04.	3.8 (±0.8) ab	3.6 (±0.8) a	6.8 (±0.8) b	3.9 (±0.6) ab	**23.04.**	7.2 (±0.3) a	6.2 (±0.3) ab	5.0 (±0.3) b	2.7 (±0.4) c
	25.04.	3.9 (±1.0) a	1.0 (±0.3) b	1.7 (±0.4) ab	1.1 (±0.2) b	**28.04.**	5.0 (±0.5) a	0.5 (±0.1) b	0.5 (±0.1) b	0.3 (±0.1) b

Tab. 2: Mean (±SE) infestation of turnip rape (TR) cultivars and oilseed rape (OSR) by larvae of *Psylliodes chrysocephala*, *Ceutorhynchus napi*, *C. pallidactylus*, *C. obstrictus*; *mean of only one location; different letters in a column indicate significant differences within one year (α≤0.05)

Year	Cultivar	*P. chrysocephala* (larvae per plant)	*C. napi* (larvae per stem)	*C. pallidactylus* (larvae per stem)	*C. obstrictus* (exit holes per 100 pods)
2009	Oase (OSR)	**2.00** (±0.26) a*	**0.02** (±0.01) a	**0.47** (±0.11) a	**1.30** (±0.36) a
	Largo (TR)	**5.72** (±0.39) b*	**0.03** (±0.03) a	**0.15** (±0.07) a	**1.08** (±0.38) a
	Prisma (TR)	**5.43** (±0.60) b*	**0.05** (±0.03) a	**0.42** (±0.19) a	**1.80** (±0.60) a
	Perko (TR)	**6.56** (±0.31) b*	**0.07** (±0.04) a	**1.02** (±0.17) b	**0.68** (±0.35) a
2010	Robust (OSR)	**1.96** (±0.42) a	**0.19** (±0.08) a	**0.86** (±0.37) a	**26.67** (±4.35) a
	Largo (TR)	**8.07** (±1.02) b	**0.09** (±0.05) a	**0.28** (±0.08) a	**5.31** (±1.61) b
	Malwira (TR)	**6.81** (±0.50) b	**0.40** (±0.07) a	**2.32** (±0.75) a	**5.64** (±1.43) b
	Perko (TR)	**5.68** (±0.77) b	**0.26** (±0.11) a	**1.48** (±0.56) a	**5.88** (±1.22) b
2011	Robust (OSR)	**2.64** (±0.55) a	**1.88** (±0.10) a	**2.95** (±0.48) a	**21.15** (±0.79) a
	Largo (TR)	**4.89** (±1.49) ab	**1.37** (±0.16) a	**2.09** (±0.29) a	**11.82** (±2.41) b
	Malwira (TR)	**4.35** (±0.70) ab	**1.83** (±0.20) a	**2.55** (±0.41) a	**10.43** (±1.62) b
	Perko (TR)	**5.95** (±0.44) b	**1.48** (±0.14) a	**2.80** (±0.34) a	**10.35** (±0.82) b

Tab. 3: Mean (±SE) infestation of oilseed rape ('Oase': 2009; 'Robust': 2010, 2011) in pure stand and in a mixture with turnip rape (95%:5% OSR/TR & 90%:10% OSR/TR) by larvae of *Psylliodes chrysocephala*, *Ceutorhynchus napi*, *C. pallidactylus*, *C. obstrictus*; different letters in a column indicate significant differences within one year (α≤0.05)

Year	Treatment	***P. chrysocephala*** (larvae per plant)	***C. napi*** (larvae per stem)	***C. pallidactylus*** (larvae per stem)	***C. obstrictus*** (exit holes per 100 pods)
	Oase 100%		0.02 (±0.01) a	0.47 (±0.11) a	1.3 (±0.4) a
2009	Oase 95%		0 02 (±0.02) a	0.29 (±0.08) a	1.3 (±0.3) a
	Oase 90%		0.04 (±0.04) a	0.24 (±0.07) a	1.1 (±0.4) a
	Robust 100%	1.96 (±0.42) a	0.19 (±0.08) a	0.86 (±0.37) a	26.7 (±4.3) a
2010	Robust 95%		0.09 (±0.04) a	0.63 (±0.31) a	22.8 (±2.7) ab
	Robust 90%	1.59 (±0.20) a	0.14 (±0.07) a	0.90 (±0.33) a	18.2 (±1.3) b
	Robust 100%	2.64 (±0.46) a	1.88 (±0.51) a	2.95 (±0.72) a	21.2 (±4.6) a
2011	Robust 95%		1.33 (±0.34) a	2.69 (±0.58) a	22.7 (±6.0) a
	Robust 90%	2.96 (±0.47) a	1.53 (±0.45) a	2.84 (±0.42) a	21.0 (±5.6) a

Tab. 4: Mean (±SE) infestation of oilseed rape (OSR) ('Oase': 2009; 'Robust': 2010, 2011) in pure stand and in mixtures with turnip rape (95%:5% OSR/TR & 90%:10% OSR/TR) by *Meligethes aeneus* during bud stage. Different letters in a row indicate significant differences (α≤0.05)

	Goettingen				Witzenhausen			
Year	Date	**OSR 100%**	**OSR 95%**	**OSR 90%**	Date	**OSR 100%**	**OSR 95%**	**OSR 90%**
	07.04.	0.8 (±0.2) a	0.7 (±0.1) a	0.8 (±0.1) a	**08.04.**	0.7 (±0.1) a	0.4 (±0.1) a	0.5 (±0.2) a
	09.04.	1.5 (±0.1) a	1.8 (±0.3) a	1.6 (±0.0) a	**11.04.**	2.2 (±0.0) a	1.7 (±0.2) a	1.8 (±0.2) a
	12.04.	3.2 (±0.1) a	3.2 (±0.3) a	3.5 (±0.2) a	**14.04.**	2.5 (±0.4) a	2.8 (±0.4) a	2.8 (±0.3) a
2009	**15.04.**	4.0 (±0.3) a	4.3 (±0.2) a	4.6 (±0.2) a	**19.04.**	1.7 (±0.2) a	2.0 (±0.4) a	2.0 (±0.4) a
	20.04.	5.1 (±0.6) a	5.0 (±0.6) a	5.2 (±0.5) a	**22.04.**	1.8 (±0.1) a	1.7 (±0.2) a	1.8 (±0.3) a
	23.04.	0.8 (±0.1) a	0.8 (±0.1) a	0.9 (±0.1) a	**25.04.**	0.8 (±0.1) a	0.9 (±0.1) a	1.0 (±0.1) a
					30.04.	0.4 (±0.1) a	0.5 (±0.1) a	0.4 (±0.1) a
	09.04.	0.2 (±0.0) a	0.2 (±0.0) a	0.3 (±0.0) a	**07.04.**	0.6 (±0.1) a	0.3 (±0.1) a	0.3 (±0.1) a
	13.04.	0.2 (±0.1) a	0.3 (±0.1) a	0.3 (±0.0) a	**13.04.**	0.4 (±0.0) a	0.3 (±0.1) a	0.3 (±0.1) a
	16.04.	0.3 (±0.0) a	0.3 (±0.1) a	0.4 (±0.0) a	**17.04.**	1.3 (±0.2) a	0.8 (±0.1) a	0.7 (±0.1) a
	19.04.	0.7 (±0.2) a	0.8 (±0.3) a	0.5 (±0.0) a	**19.04.**	1.4 (±0.1) a	0.9 (±0.2) a	0.9 (±0.1) a
	22.04.	0.7 (±0.1) a	0.8 (±0.4) a	0.8 (±0.2) a	**22.04.**	1.1 (±0.1) a	0.9 (±0.1) a	0.9 (±0.1) a
2010	**25.04.**	5.2 (±0.7) a	4.1 (±0.6) a	3.9 (±0.3) a	**24.04.**	2.1 (±0.0) a	2.0 (±0.3) a	2.2 (±0.2) a
	29.04.	4.8 (±0.3) a	4.8 (±0.5) a	5.2 (±0.1) a	**27.04.**	5.2 (±0.3) a	5.1 (±0.5) a	4.7 (±0.3) a
	01.05.	6.3 (±0.5) a	6.5 (±0.2) a	6.2 (±0.2) a	**30.04.**	7.2 (±0.8) a	7.1 (±1.0) a	6.5 (±0.4) a
	05.05.	3.8 (±0.4) a	3.7 (±0.2) a	3.3 (±0.3) a	**03.05.**	7.1 (±0.7) a	6.0 (±0.5) a	5.8 (±0.7) a
					08.05.	4.5 (±0.5) a	3.2 (±0.4) a	3.5 (±0.3) a
	09.04.	0.3 (±0.1) a	0.4 (±0.2) a	0.4 (±0.1) a	**08.04.**	0.6 (±0.1) a	0.6 (±0.0) a	0.6 (±0.1) a
	13.04.	0.4 (±0.1) a	0.5 (±0.1) a	0.4 (±0.1) a	**10.04.**	1.4 (±0.2) a	1.0 (±0.1) a	1.0 (±0.2) a
	16.04.	1.3 (±0.4) a	1.7 (±0.3) a	1.3 (±0.3) a	**14.04.**	1.6 (±0.2) a	1.4 (±0.1) a	1.5 (±0.3) a
2011	**19.04.**	2.0 (±0.5) a	2.0 (±0.3) a	1.4 (±0.2) a	**17.04.**	1.9 (±0.1) a	1.9 (±0.2) a	1.8 (±0.2) a
	22.04.	3.8 (±0.8) a	3.8 (±0.3) a	2.7 (±0.2) a	**20.04.**	5.4 (±0.3) a	5.2 (±0.2) a	5.6 (±0.2) a
	25.04.	3.9 (±1.0) a	4.3 (±0.9) a	3.1 (±0.5) a	**23.04.**	7.2 (±0.3) a	8.2 (±0.5) a	8.3 (±0.4) a
					28.04.	5.0 (±0.5) a	4.2 (±0.3) a	4.5 (±0.3) a

Tab. 5: Mean (±SE) damage caused by stem mining larvae (stem injury index), by pollen beetles (bud loss index) and seed yield of oilseed rape (OSR) (Oase: 2009; Robust: 2010, 2011) in pure stand and in mixtures with turnip rape (95%:5% OSR/TR & 90%:10% OSR/TR); *mean of only one location. Different letters in a column indicate significant differences in within one year at α≤0.05

Year	Treatment	Stem injury index	Bud loss index	Seed yield (dt/ha)
	Oase	0.03 (±0.00) a	0.22 (±0.04) a	19.1 (±1.8) a
2009	Oase 95%	0.02 (±0.00) a	0.15 (±0.01) a	17.0 (±2.0) a
	Oase 90%	0.02 (±0.00) a	0.15 (±0.01) a	17.9 (±1.7) a
	Robust	0.06 (±0.01) a	0.35 (±0.04) a	13.6 (±1.5) a
2010	Robust 95%	0.05 (±0.01) a	0.36 (±0.03) a	14.6 (±1.7) a
	Robust 90%	0.07 (±0.02) a	0.35 (±0.03) a	15.0 (±1.4) a
	Robust	0.17 (±0.04) a	0.56 (±0.05) a	14.7 (±1.2) a*
2011	Robust 95%	0.14 (±0.02) a	0.51 (±0.04) a	15.6 (±1.1) a*
	Robust 90%	0.17 (±0.03) a	0.50 (±0.05) a	14.6 (±0.7) a*

Stem injury index and bud loss index were on a relatively low level in 2009 and increased considerably in 2010 and 2011 (Tab. 5). In all years, irrespective of the level of damage, there was no significant difference between stem injury index or bud loss index of oilseed rape plants in pure stand and in mixture with turnip rape (Tab. 5). The seed yield was not significantly different between pure stands of oilseed rape and the two mixed cropping treatments in any experiment (Tab. 5).

Discussion

In our experiments the tested cultivars of turnip rape were clearly preferred rape by cabbage stem flea beetle and pollen beetle to oilseed. Similarly, in earlier studies both species were reported to prefer turnip rape to oilseed rape for oviposition (Büchi, 1990; 1995; Cook, 2003; Barari et al., 2005) while rape stem weevil, cabbage stem weevil and cabbage seed weevil were not found to show higher levels of infestation on turnip rape than on oilseed rape (Büchi, 1990; 1995; Cook, 2003; Kalischuk and Dosdall, 2004). When individual cultivars of turnip rape and oilseed rape were assessed for their susceptibility to oilseed rape pests the results were inconsistent. In zhe UK cabbage stem flea beetle, pollen beetle and brassica pod midge showed a significant preference for individual cultivars of oilseed rape (Williams, 1989; Lambdon et al., 1998). In studies from Sweden and Canada, the oilseed rape pests *Phyllotreta undulata* and *Mamestra configurata* did not discriminate between cultivars and lines of oilseed rape or turnip rape (Ekbom, 1995; Ulmer et al., 2001). In our experiments, only pollen beetles responded differentially to the tested turnip rape cultivars and preferred the cultivar 'Malwira' significantly.

Finch and Collier (2000) stated that host selection by herbivorous insects is a sequence of three steps. First, host location is driven by olfactory stimuli derived from the host plant. Second, host selection over short distances is driven by visual cues. Third, non-volatile plant chemicals are used for acceptance of host plant after landing (Williams and Bartlet, 1993; Finch and Collier, 2000). The plant factors responsible for the different attractiveness and larval infestation of turnip rape and oilseed rape or different turnip rape cultivars could not be determined in our experiment.

In earlier studies volatiles released from turnip rape were more attractive to insect pests than volatiles of oilseed rape (Cook et al., 2007). When the odour of turnip rape and oilseed rape was offered to pollen beetles in an olfactometer, significantly higher numbers preferred the turnip rape odour (Cook et al., 2007). Specific Brassica volatiles that are derived from degradation of aliphatic glucosinolates are also known to attract cabbage stem flea beetle and cabbage seed weevil (Bartlet et al., 1993; Bartlet et al., 1997). Therefore the glucosinolate composition might be an important factor for host finding and acceptance of specialist insects on oilseed rape (Bartlet et al., 1999; Cook et al., 2007). The identity of volatiles emitted from the tested cultivars could not be analysed in our experiments.

Visual cues have been also reported in previous studies as a major factor for host selection by herbivorous insects (Schoonhoven et al. 2005). For pollen beetle, the faster development and earlier start of flowering of turnip rape may be responsible for their preference as has been reported in the literature (Williams and Free, 1978; Nilsson, 2004; Frearson et al., 2005). Later in the season, higher numbers of pollen beetles on oilseed rape compared to turnip rape may be explained by a change of the preferred host cues (Frearson et al., 2005). At this stage, buds suitable for oviposition are more important for pollen feeding than flowers (Frearson et al., 2005). Cabbage seed weevil also was found to prefer flowering host plants during immigration in spring (Dmoch, 1965). When flowering of these plants is completed, the weevils move to other plants that are still in flower (Dmoch, 1965). Therefore, the different phenologies of oilseed rape and turnip rape may be responsible for the higher infestation of oilseed rape compared to turnip rape.

In our experiments the turnip rape plants were growing more vigorously than oilseed rape plants (data not shown), which possibly had major influence on host plant selection of the tested species. As stated by the “plant vigor hypothesis”, many herbivorous insect species prefer more vigorous host plants to backward plants (Price, 1991; Cornelissen et al., 2008). This was reported e.g. for cabbage root flies (*Delia* spp.), common pests on oilseed rape in Canada, which prefer taller plants for oviposition (Dosdall et al., 1996). The length of the main shoot is an important morphological cue for host selection by ovipositing rape stem weevils

(Büchi, 1996). Furthermore, the size of canopy and the plant coverage of the soil might play a major role especially for the cabbage stem flea beetle. Nuss (2004) suggested that a dense canopy leads to a higher soil moisture and provides better conditions for oviposition of cabbage stem flea beetle. Generally, there is little information on the effect of plant architecture on host selection by oilseed rape pests. Therefore in further experiments the leaf area index and plant architecture parameters at the time of pest immigration should be taken into account.

When pest insects have arrived on the plant surface, chemical plant substances, such as glucosinolates and sugars are important factors for host acceptance and stimulation of feeding and oviposition (Finch and Collier, 2000; Schoonhoven et al., 2005). Furthermore, morphological structures such as trichomes or epicuticular waxes are important factors for the acceptance of a host plant (Hoxie et al., 1975; Bodnaryk, 1992; Schoonhoven et al., 2005). Earlier studies have shown that trichome density and thickness of epicuticular wax layer differed between oilseed rape and turnip rape and within cultivars of oilseed rape (Bodnaryk, 1992; Lambdon et al., 1998; Soroka et al., 2011). On plants providing a high trichome density and a thick layer of epicuticular waxes the feeding of flea beetles was significantly reduced (Bodnaryk, 1992; Lambdon et al., 1998; Soroka et al., 2011).

A highly attractive trap crop was supposed to maintain the pest insects and to protect the main crop from pest damage (Hokkanen, 1991; Cook, 2002). In our experiments mixed cropping of turnip rape and oilseed rape did not reduce the infestation of oilseed rape plants. Even a high proportion of turnip rape (10%) had no effect on the level of pest infestation on oilseed rape. In earlier studies seed mixtures of oilseed rape and turnip rape (98:2 oilseed rape : turnip rape) have shown potential to reduce pollen beetle infestation on oilseed rape plants and to reduce the need for insecticide application (Büchi, 1990; Nilsson, 2004). For cabbage stem flea beetle not only the oviposition preference of females may play a role for different levels of infestation of turnip rape and oilseed rape in the mixture. Previous studies have shown that larvae are able to migrate from plant to plant towards preferred plant volatiles (Queinnec, 1967).

Our results suggest that 'Malwira' may increase the efficacy of the trap crop for control of pollen beetles in oilseed rape fields but not for other pest species. Due to the high content of erucic acid and glucosinolates in the seeds, trap cropping by growing the cultivar 'Malwira' as a trap crop would have a negative impact on the quality of the harvested seeds (Bell, 1995; Uppström, 1995). This might reduce the implementation of this strategy in practical farming.

In general, in order to lure insect pests away from oilseed rape, the trap crop should grow up more vigorously, develop faster in spring and emit a highly attractive blend of volatiles (Cook et al., 2007). Moreover, the leaves of a trap crop should have low amounts of epicuticular waxes and other feeding deterrents (Shelton and Badenes-Perez, 2006; Cook et al., 2007). They should contain high amounts of feeding stimulats to ensure that the plant is accepted at the insects' first landing (Finch and Collier, 2000; Cook et al., 2007). Under controlled climate conditions, adult cabbage stem flea beetles significantly preferred turnip rape for feeding when it was offered together with oilseed rape (Chapter III). Leaves of the host plants were analysed for sugars, phenolics and glucosinolates, as these chemicals are known to affect feeding of herbivorous insects. However, there was no correlation between the tested plant cues and the feeding preference of cabbage stem flea beetle.

References

Alford DV, Nilsson C & Ulber B (2003) Insect pests of oilseed rape crops. In: Biocontrol of oilseed rape pests (ed. DV Alford). Blackwell Science Ltd., Oxford, pp. 9-42.

Barari H, Cook SM, Clark SJ & Williams IH (2005) Effect of a turnip rape (*Brassica rapa*) trap crop on stem-mining pests and their parasitoids in winter oilseed rape (*Brassica napus*). Biocontrol 50: 69-86.

Bartlet E, Blight MM, Hick AJ & Williams IH (1993) The responses of the cabbage seed weevil (*Ceutorhynchus assimilis*) to the odour of oilseed rape (*Brassica napus*) and to some volatile isothiocyanates. Entomologia Experimentalis et Applicata 68: 295-302.

Bartlet E, Blight MM, Lane P & Williams IH (1997) The responses of the cabbage seed weevil *Ceutorhynchus assimilis* to volatile compounds from oilseed rape in a linear track olfactometer. Entomologia Experimentalis et Applicata 85: 257-262.

Bartlet E, Williams IH & Pickett JA (1999) The ideal glucosinolate profile for pest resistance in oilseed rape. IOBC/wprs Bulletin 22 (10): 13-17.

Bell JM (1995) Meal and by-product utilization in animal nutrition. In: Brassica oilseeds - production and utilization (ed. DS Kimber & DI McGregor). CAB International, Wallingford, pp. 301-338.

Bodnaryk RP (1992) Leaf epicuticular wax, an antixenotic factor in Brassicaceae that affects the rate and pattern of feeding of flea beetles, *Phyllotreta cruciferae* (Goeze). Canadian Journal of Plant Science 72: 1295-1303.

Büchi R (1990) Investigation on the use of turnip rape as a trap plant to control oilseed rape pests. IOBC/wprs Bulletin 13 (4): 32-39.

Büchi R (1995) Combination of trap plants (*Brassica rapa var. silvestris*) and insecticide use to control rape pests. IOBC/wprs Bulletin 18 (4): 102-119.

Büchi R (1996) Eiablage des Rapsstängelrüsslers *Ceutorhynchus napi* Gyll., in Abhängigkeit der Stängellänge bei verschiedenen Rapssorten. Anzeiger für Schädlingskunde, Pflanzenschutz, Umweltschutz 69: 136-139.

Büchs W (2009) Fangpflanzen (trap crop) als Methode zur Kontrolle tierischer Schädlinge im ökologischen Rapsanbau. Mitteilungen der Deutschen Gesellschaft für Allgemeine und Angewandte Entomologie 17: 241-246.

Carcamo HA, Dunn R, Dosdall LM & Olfert O (2007) Managing cabbage seedpod weevil in canola using a trap crop - A commercial field scale study in western Canada. Crop Protection 26: 1325-1334.

Cook SM, L. E. Smart, H. B. Rasmussen, E. Bartlet, J. L. Martin, D. A. Murray, N. P. Watts, I. H. Williams (2003) Push-Pull strategies to reduce insecticide input to oilseed rape (*Brassica napus*): Potential of low alkenyl glucosinolate oilseed rape varieties (push!), and turnip rape (*Brassica rapa*) trap crops (pull!): 11th International Rapeseed Congress Copenhagen, pp. 1018-1021.

Cook SM, L. E. Smart, R. J. P. Potting, E. Bartlet, J. L. Martin, D. A. Murray, N. P. Watts, I. H. Williams (2002) Turnip rape (*Brassica rapa*) as a trap crop to protect oilseed rape (*Brassica napus*) from infestation by insect pests:

potential and mechanisms of action The BCPC Conference: Pests & Diseases 2002 2: 569-574.

Cook SM, Rasmussen HB, Birkett MA, Murray DA, Pye BJ, Watts NP & Williams IH (2007) Behavioural and chemical ecology underlying the success of turnip rape (*Brassica rapa*) trap crops in protecting oilseed rape (*Brassica napus*) from the pollen beetle (*Meligethes aeneus*). Arthropod-Plant Interactions 1: 57-67.

Cook SM, Smart LE, Martin JL, Murray DA, Watts NP & Williams IH (2006) Exploitation of host plant preferences in pest management strategies for oilseed rape (*Brassica napus*). Entomologia Experimentalis et Applicata 119: 221-229.

Cook SM, Watts NP, Hunter F, Smart LE & Williams IH (2004) Effects of a turnip rape trap crop on the spatial distribution of *Meligethes aeneus* and *Ceutorhynchus assimilis* in oilseed rape. IOBC/wprs Bulletin 27 (10): 199-206

Cornelissen T, Fernandes GW & Vasconcellos-Neto J (2008) Size does matter: variation in herbivory between and within plants and the plant vigor hypothesis. Oikos 117: 1121-1130.

Dmoch J (1965) The dynamics of population of the cabbage seedpod weevil (*Ceutorhynchus assimilis* Payk.) and the development of winter rape. Part 1. Ekologia Polska Seria A 8: 249-287.

Dosdall LM, Herbut MJ, Cowle NT & Micklich TM (1996) The effect of seeding date and plant density on infestations of root maggots, *Delia* spp, (Diptera: Anthomyiidae), in canola. Canadian Journal of Plant Science 76: 169-177.

Eickermann M, Ulber B, Vidal S 2010. Resynthesized lines and cultivars of B*rassica napus* L. provide sources of resistance to the cabbage stem weevil (*Ceutorhynchus pallidactylus* (Marsh.)). Bulletin of Entomological Research 110, 1-8.

Ekbom B (1995) Flea beetle (*Phyllotreta undulata*) feeding preference. IOBC/wprs Bulletin 18 (4): 88-94.

Evans KA & Allen-Williams LJ (1994) Laboratory and field response of the pollen beetle, *Meligethes aeneus*, to the odour of oilseed rape. Physiological Entomology 19: 285-290.

Finch S & Collier RH (2000) Host-plant selection by insects - a theory based on 'appropriate/inappropriate landings' by pest insects of cruciferous plants. Entomologia Experimentalis et Applicata 96: 91-102.

Frearson DJT, Ferguson AW, Campbell JM & Williams IH (2005) The spatial dynamics of pollen beetles in relation to inflorescence growth stage of oilseed rape: implications for trap crop strategies. Entomologia Experimentalis et Applicata 116: 21-29.

Fritzsche R (1957) Zur Biologie und Ökologie der Rapsschädlinge aus der Gattung *Meligethes*. Journal of Applied Entomology 40: 222-280.

Günthart E (1949) Beiträge zur Lebensweise und Bekämpfung von *Ceutorrhynchus quadridens* PANZ. und *Ceutorrhynchus napi* GYLL. Mitteilungen der Schweizerischen Entomologischen Gesellschaft 22: 443-591.

Hokkanen HMT (1991) Trap cropping in pest management. Annual Review of Entomology 36: 119-138.

Hopkins RJ, van Dam NM & van Loon JJA (2009) Role of glucosinolates in insect-plant relationships and multitrophic interactions. Annual Review of Entomology 54: 57-83.

Hoxie RP, Wellso SG & Webster JA (1975) Cereal leaf beetle response to wheat trichome length and density. Environmental Entomology 4: 365-370.

Kalischuk AR & Dosdall LM (2004) Susceptibilities of seven Brassicaceae species to infestation by the cabbage seedpod weevil (Coleoptera: Curculionidae). Canadian Entomologist 136: 265-276.

Kaufmann O (1941) Zur Biologie des Rapserdflohes (*Psylliodes chrysocephala* L.). Zeitschrift für Pflanzenkrankheiten (Pflanzenpathologie) und Pflanzenschutz 51: 305-324.

Lambdon PW, Hassall M & Mithen R (1998) Feeding preferences of woodpigeons and flea-beetles for oilseed rape and turnip rape. Annals of Applied Biology 133: 313-328.

Lancashire PD, Bleiholder H, Vandenboom T, Langeluddeke P, Stauss R, Weber E & Witzenberger A (1991) A uniform decimal code for growth-stages of crops and weeds. Annals of Applied Biology 119: 561-601.

Nilsson C (2004) Trap plants to avoid insecticide application against pollen beetles in oilseed rape. IOBC/wprs Bulletin 27 (10): 215-221.

Nilsson C (2010) Impact of soil tillage on parasitoids of oilseed rape pests. In: Biocontrol-based integrated management of oilseed rape pests (ed. IH Williams). Springer Science+Business Media B.V., Dordrecht, Heidelberg, London, New York, pp. 305-312.

Nuss H (2004) Einfluss der Pflanzendichte und -architektur auf Abundanz und innerpflanzliche Verteilung stängelminierender Schadinsekten in Winterraps. PhD thesis, Georg-August-Universität Göttingen.

Price PW (1991) The plant vigor hypothesis and herbivore attack. Oikos 62: 244-251.

Queinnec Y (1967) Étude des fakteurs psychophysiologiques permettant la découverte de la plante-hote par les larves néonates de l'altise d'hiver du colza (*Psylliodes chysocephala* L.). Annales des Épiphyties 18: 27-74.

Schoonhoven LM, van Loon JJA & Dicke M (2005) Insect-plant biology. Oxford University Press, Oxford.

Shelton AM & Badenes-Perez E (2006) Concepts and applications of trap cropping in pest management. Annual Review of Entomology 51: 285-308.

Soroka JJ, Holowachuk JM, Gruber MY & Grenkow LF (2011) Feeding by flea beetles (Coleoptera: Chrysomelidae; *Phyllotreta* spp.) is decreased on canola (*Brassica napus*) seedlings with increased trichome density. Journal of Economic Entomology 104: 125-136

Ulmer B, Gillott C & Erlandson M (2001) Feeding preferences, growth, and development of *Mamestra configurata* (Lepidoptera: Noctuidae) on Brassicaceae. Canadian Entomologist 133: 509-519.

Uppström B (1995) Seed chemistry. In: Brassica oilseeds production and utilization (ed. DS Kimber & DI McGregor). CAB International, Wallingford, pp. 217-242.

Williams IH (1989) Pest incidence on single low and double low oilseed rape. Aspects of Applied Biology 23: 277-286.

Williams IH (2004) Advances in insect pest management of oilseed rape in Europe. In: Insect pest management (ed. AR Horowitz & I Ishaaya). Springer-Verlag, Berlin, Heidelberg, pp. 181-208.

Williams IH (2010) The major insect pests of oilseed rape in Europe and their management: an overview. In: Biocontrol-based integrated management of oilseed rape pests (ed. IH Williams). Springer Science+Business B.V., Dordrecht, Heidelberg, London, New York, pp. 1-43.

Williams IH & Bartlet E (1993) Host plant affecting the feeding of cabbage stem flea beetle (*Psylliodes chrysocephala* L.) on oilseed rape. IOBC/wprs Bulletin 16 (9): 202-209.

Williams IH & Free JB (1978) The feeding and mating behaviour of pollen beetles (*Meligethes aeneus* Fab.) and seed weevils (*Ceutorhynchus assimilis* Payk.) on oil-seed rape (*Brassica napus* L.). Journal of Agricultural Science 91: 453-459.

Winfield AL (1992) Management of oilseed rape pests in Europe. Agricultural Zoology Reviews 5: 51-95.

Chapter V: Management of insect pests on winter oilseed rape by perimeter strips of turnip rape

Abstract

Insect pests on oilseed rape (*Brassica napus* L.) at present are commonly controlled by the application of insecticides. Due to the emerging resistance in many pests against frequently applied active compounds and the negative impact of insecticides on the environment, alternative strategies are needed to boost integrated pest management. In a two-year field study, a trap cropping system including turnip rape (*Brassica rapa* L. var. *silvetris*) as trap crop was evaluated. Perimeter strips with alternating plots of oilseed rape and turnip rape were sown along the edges of oilseed rape crops on two opposite sites. During the growing season the abundance of cabbage stem flea beetle (*Psylliodes chrysocephala* L.), rape stem weevil (*Ceutorhynchus napi* Gyll.), cabbage stem weevil (*Ceutorhynchus pallidactylus* Marsh.), pollen beetle (*Meligethes aeneus* Fab.) and cabbage seed weevil (*Ceutorhynchus obstrictus* Marsh.) was determined in the perimeter strip plots and in the adjacent main crop at distances of 10m and 30m apart from the perimeter strip. The seed yield of oilseed rape in the perimeter strip and the main crop was assessed as well. Within the perimeter strips cabbage stem flea beetles and pollen beetles significantly preferred turnip rape over oilseed rape, whereas the stem weevils did not show a preference for one of the plant species. Cabbage seed weevil infestation in the perimeter strip was higher on oilseed rape than on turnip rape. Pollen beetle abundance on turnip rape and oilseed rape was mainly affected by plant phenology: during the early immigration phase significantly more pollen beetles colonized turnip rape than oilseed rape, but this difference was less obvious later in the season. The infestation level of the oilseed rape main crop by all pest species studied was not affected significantly by the turnip rape perimeter strip, neither at a distance of 10m nor at a distance of 30m apart from the perimeter strip. These results suggest that turnip rape trap crop strips are not appropriate for managing insect pests on oilseed rape.

Introduction

The acreage of oilseed rape (*Brassica napus* L.) in Europe has substantially increased during the last decades, extending the need for crop protection inputs (Alford, 2003). One major problem of oilseed rape production in Europe is the infestation by insect pests, predominantly by species of the order Coleoptera (Alford et al., 2003; Williams, 2010). The cabbage stem flea beetle (*Psylliodes chrysocephala* L.) colonizes the oilseed rape crop in September shortly after plant emergence (Williams, 2010). Adult beetles feed on leaf lamina; this damage is less important than the damage caused by larvae (Williams, 2010). In early spring, two stem-boring weevil species, *Ceutorhynchus napi* Gyll. and *C. pallidactylus* Marsh., invade the oilseed rape fields. During stem elongation, females lay their eggs into stems and petioles of oilseed rape plants, respectively (Lerin, 1993; Williams, 2010). Pollen beetle (*Meligethes aeneus* Fab.) another important insect pest colonizing the oilseed rape fields during the bud stage in spring is (Fritzsche, 1957). To get access to the pollen, the beetles feed through the bud walls, thereby damaging the ovaries and inducing bud losses (Williams & Free, 1978). Later in the season the developing pods are exposed to infestation by cabbage seed weevil (*Ceutorhynchus obstrictus* Marsh.) and cabbage pod midge (*Dasineura brassicae* Winn.) (Williams, 2010). Larvae of both species develop within the pods and can reduce seed yields significantly (Williams, 2010). At present, control of all pest species strongly relies on the use of insecticides. These chemicals can have negative effects on the environment and on beneficial insects (Williams, 2004; Bommarco et al., 2011). Recently, insecticide resistant populations of the cabbage stem flea beetle and the pollen beetle have been reported in Germany (Heimbach & Müller, 2011; Slater et al., 2011). Overall, integrated pest management strategies are needed which have potential to reduce the need for insecticide application to a minimum.

The use of trap crops is one promising option for integrated pest management (Hokkanen, 1991). The effects of trap crops have been studied in various oilseed rape growing countries (Hokkanen et al., 1986; Büchi, 1990; Nilsson, 2004; Barari et al., 2005; Carcamo et al., 2007). A trap crop is defined as a plant species that is more attractive for insect pests than the targeted crop (Hokkanen, 1991). When both plant species are grown close to each other, the insect pests are lured away from the targeted crop and aggregated on the trap crop (Hokkanen, 1991), thereby protecting the targeted crop from damaging infestation levels of pests (Hokkanen, 1991). The most promising plant species for trap cropping in oilseed rape is turnip rape (*Brassica rapa* L . var. *silvestris*) (Cook et al., 2002): this species is preferred by many specialist insect pests to oilseed rape (Büchi, 1995; Lambdon et al., 1998; Cook et al., 2006; Cook et al., 2007). The higher attractiveness of turnip rape compared to oilseed rape results from different olfactory and visual cues emitted by these plant species (Büchi, 1990; Giamoustaris & Mithen, 1996; Cook et al., 2007). Previous studies reported that during the early infestation period the edges of oilseed rape fields are more severely infested by many insect pests than the interior field (Attah & Lawton, 1984). At later colonization phases, most insect pests spread in patches over the whole oilseed rape field (Thioulouse, 1987; Ferguson et al., 2003). In order to concentrate and maintain the insect pests at the edges of the field and to reduce their migration into the main crop a turnip rape trap crop can be grown in perimeter strips (Cook et al., 2004). In this investigation, the efficacy of this strategy was assessed under farm conditions in Central Germany.

Material and Methods

Experimental setup of field trials

Field experiments were conducted in 2008/09 and 2009/10, with two replicated experiments at different sites in each year. Fields were located on organically managed farms in southern Lower Saxony (LS) (Wiebrechtshausen in 2009, Ebergoetzen in 2010) and in northern Hesse (HE) (Neu-Eichenberg) 2009 + 2010. The field size ranged from two to five hectares; the shape of the fields was almost rectangular. Along the edges, the main oilseed rape crop was bordered by eight alternating plots of turnip rape and oilseed rape, respectively (plot size: 60m long, 6m wide). Depending on the length of the field, these plots were established either on one site or on two opposite sites. The cultivar 'Largo' was chosen for the turnip rape plots. In the oilseed rape perimeter plots the same cultivar as in the main crop was grown. As the farmers used different cultivars of oilseed rape, these differed between experiments. On the fields in LS, the oilseed rape cultivars

'Robust' and 'Alkido' were grown in 2009 and 2010, respectively, while the oilseed rape cultivars 'Oase' and 'Robust' were grown in 2009 and in 2010 in HE, respectively. Due to severe plant mortality caused by frost in January, no assessments were possible on the experimental field in LS in 2010.

Data assessement

The infestation of turnip rape and oilseed rape by insect pests and the damage caused by these pests was assessed in the perimeter plots and in the adjacent main crop at distances of 10m and 30m to the perimeter plots. To avoid edge effects of neighbouring plots insect and plant samples were collected within the central 20m of each plot, 20m apart from the next plot. From the beginning of the bud stage to mid flowering the growth stages (Lancashire et al., 1991) of turnip and oilseed rape were recorded twice a week. On two occasions, 20 plants were randomly collected from each plot. The first sampling was in March when oilseed rape and turnip rape were in BBCH 29. The number of cabbage stem flea beetle larvae was assessed by dissecting leaves and shoots. The second sampling was in May, in BBCH 67, towards the end of flowering. At this date larvae of cabbage stem weevil and rape stem weevil were dissected from stems of oilseed rape and turnip rape and counted. At the same date the plant height and length of the feeding gallery were measured to calculate the stem injury index (Eickermann et al. 2010):

$$\text{Stem injury index} = \text{length of feeding gallery} * \text{stem length}^{-1}$$

The abundance of pollen beetle was assessed from early bud stage (BBCH 51) to mid flowering (BBCH 65). In each sampling the pollen beetle infestation was assessed on 50 randomly selected plants. Beetles were beaten from the inflorescences into a plastic tray (26.5cm * 33.5cm) and counted. To assess the damage caused by pollen beetles, the bud loss index was calculated, based on the number of empty stalks and the total number of pods on the terminal and 2nd lateral racemes of 20 plants in each plot:

Bud loss index = empty stalks * (empty stalks + pods)$^{-1}$

Cabbage seed weevil infestation was assessed by counting the number of exit holes caused by emigrating larvae. The holes were counted on pods of the terminal and 2nd lateral raceme of 20 plants or from 200 randomly chosen pods of the terminal and 2nd lateral racemes of 20 plants. To measure the seed yield, subplots of 15m² to 22,5m² were harvested at growth stage BBCH 97 by using a parcel combine.

Data analysis

In 2009, when data were obtained from experiments at both locations ANCOVA was conducted with the location set as covariable. In 2010, data of the experiment in LS were analysed by using ANOVA. ANCOVA and ANOVA were followed by Tukey HSD test used as post-hoc-test. At $p \leq 0.05$ differences between means were treated as significant. Indices were arcsinus transformed before statistical analyses. Data sets showing inhomogenous variances were log(x+1) transformed before analysis. For all statistical analyses Statistica Version 9.1 (StatSoft Inc.) was used.

Results

In the experiments LS 2009 and in HE 2010, turnip rape started flowering eleven and six days earlier than oilseed rape, respectively. At HE 2009 the plant phenology of turnip rape and oilseed rape between early bud and mid-flowering stage was only little different.

In the perimeter plots infestation of turnip rape plants by cabbage stem flea beetle larvae was significantly higher than infestation of oilseed rape plants (Tab. 1). In both years and at both locations, plant infestation by rape stem weevil larvae and cabbage stem weevil larvae was on a low level. The number of larvae within stems did not differ significantly between oilseed rape and turnip rape (Tab. 1). The stem injury index in the perimeter plots was significantly higher in turnip rape plants than in oilseed rape plants in both years (Tab. 2). In the experiment HE in 2009, when the phenology of turnip rape and oilseed rape showed little difference, the infestation level by pollen beetle in the early bud stage was similar in both plant species. At the beginning of flowering in the perimeter strip, numbers of pollen beetles on the inflorescences of turnip rape plants were significantly higher than the numbers on oilseed rape (Tab. 3). In the following growth stages pollen beetle numbers were not different between both plant species. In LS in 2009, during the early beetle immigration phase, the number of pollen beetles on turnip rape in the perimeter strips was significantly higher than the number on oilseed rape. At the end of the bud stage of oilseed rape in the perimeter strip, the differences between beetle infestation of turnip rape and oilseed rape became less obvious. During flowering of both species even higher numbers of beetles were found on oilseed rape.

Infestation by pollen beetle increased substantially from 2009 to 2010, with a maximum of 4.6 and 16.8 beetles per terminal raceme, respectively. In 2010 in the perimeter strips, the number of beetles on turnip rape was significantly higher than on oilseed rape until turnip rape plants started flowering (Tab. 3). At mid-flowering of turnip rape and start of flowering of oilseed rape, beetle numbers on both species were no longer different. No significant difference was detected between the bud loss index of turnip rape and oilseed rape in the perimeter strip (Tab. 2). Infestation by cabbage seed weevil larvae of turnip rape pods was significantly lower than on oilseed rape pods in the perimeter strip (Tab.1).

Despite the differences between the levels of infestation of both Brassica species at field edges, there was no significant effect of perimeter strips on the levels of infestation by major pests in the main crop. The damage due to rape stem weevil, cabbage stem weevil and pollen beetle in the main crop, as assessed by the stem injury index and the bud loss index, was not reduced by the turnip rape perimeter strips compared to the oilseed rape perimeter strips.

In both years the seed yield of turnip rape was significantly lower than the seed yield of the main crop. The seed yield of the main oilseed rape crop did not differ at 10m or at 30m distance to the turnip rape or oilseed rape perimeter plots (Tab. 2).

Tab. 1: Mean (±SE) infestation of turnip rape (TR) and oilseed rape (OSR) perimeter strips (PS) and in the adjacent main crop (MC) in distances of 10m and 30m to perimeter strips by larvae of *P. chrysocephala*, *C. napi*, *C. pallidactylus*, *C. obstrictus*; *mean of two locations; different letters in a column indicate significant differences within one year at $p \leq 0.05$

Year	Treatment	***P. chrysocephala*** (larvae per plant)	***C. napi*** (larvae per stem)	***C. pallidactylus*** (larvae per stem)	***C. obstrictus*** (exit holes per 100 pods)
2009	**TR-PS**	6.2 (±0.4) a	0.0 (±0.0) a*	0.2 (±0.1) a*	1.1 (±0.2) a*
	OSR-PS	2.3 (±0.4) b	0.0 (±0.0) a*	0.8 (±0.4) ab*	0.8 (±0.2) a*
	MC 10m to TR-PS	2.7 (±0.1) b	0.0 (±0.0) a*	1.1 (±0.3) ab*	1.6 (±0.5) a*
	MC 10m to OSR-PS	2.6 (±0.3) b	0.0 (±0.0) a*	0.9 (±0.4) ab*	1.8 (±0.3) a*
	MC 30m to TR-PS	2.4 (±0.2) b	0.0 (±0.0) a*	1.5 (±0.6) b*	1.9 (±0.4) a*
	MC 30m to OSR-PS	2.4 (±0.2) b	0.0 (±0.0) a*	1.2 (±0.3) ab*	1.5 (±0.2) a*
2010	**TR-PS**	3.9 (±0.8) a	0.2 (±0.1) ab	0.7 (±0.2) a	12.4 (±3.9) a
	OSR-PS	1.9 (±1.0) a	0.1 (±0.0) ab	0.5 (±0.3) a	45.2 (±5.6) b
	MC 10m to TR-PS	1.7 (±0.4) a	0.4 (±0.1) a	0.2 (±0.2) a	47.0 (±3.8) b
	MC 10m to OSR-PS	1.3 (±0.4) a	0.1 (±0.1) ab	0.2 (±0.2) a	60.8 (±6.6) b
	MC 30m to TR-PS	2.0 (±0.4) a	0.1 (±0.1) ab	0.3 (±0.2) a	56.6 (±6.3) b
	MC 30m to OSR-PS	2.4 (±0.5) a	0.1 (±0.1) b	0.1 (±0.1) a	56.2 (±8.0) b

Tab. 2: Mean (±SE) stem injury index (damage caused by stem mining larvae), bud loss index (damage caused by pollen beetle) and seed yield of turnip rape (TR) and oilseed rape (OSR) perimeter strips (PS) and in the adjacent main crop (MC) in distances of 10m and 30m to perimeter strips; *mean of two locations; different letters in a column indicate significant differences within one year at p≤0.05

Year	Treatment	**Stem injury index**	**Bud loss index**	**Seed yield** (dt/ha)
2009	**TR-PS**	0.16 (±0.03) a*	0.23 (±0.01) a*	8.2 (±0.7) a
	OSR-PS	0.05 (±0.01) b*	0.22 (±0.03) a*	17.7 (±3.3) b
	MC 10m to TR-PS	0.05 (±0.01) b*	0.25 (±0.03) a*	21.2 (±1.0) b
	MC 10m to OSR-PS	0.04 (±0.01) b*	0.22 (±0.03) a*	19.6 (±3.3) b
	MC 30m to TR-PS	0.06 (±0.01) b*	0.21 (±0.03) a*	18.8 (±1.6) b
	MC 30m to OSR-PS	0.05 (±0.01) b*	0.23 (±0.03) a*	20.7 (±2.2) b
2010	**TR-PS**	0.20 (±0.04) a	0.33 (±0.02) a	4.5 (±2.1) a
	OSR-PS	0.05 (±0.02) b	0.42 (±0.03) a	9.1 (±1.7) a
	MC 10m to TR-PS	0.05 (±0.02) b	0.42 (±0.04) a	9.7 (±2.5) a
	MC 10m to OSR-PS	0.03 (±0.00) b	0.41 (±0.03) a	9.5 (±1.0) a
	MC 30m to TR-PS	0.03 (±0.01) b	0.41 (±0.02) a	12.0 (±4.6) a
	MC 30m to OSR-PS	0.03 (±0.00) b	0.48 (±0.03) a	9.4 (±2.0) a

Tab. 3: Mean (±SE) number of pollen beetle per inflorescence and growth stage of turnip rape (TR) and oilseed rape (OSR) in perimeter strips (PS) and in the adjacent main crop (MC) at distances of 10m and 30m to perimeter strips; fields were located in Hesse (HE) and Lower Saxony (LS). Different letters in a row indicate significant differences at $p \leq 0.05$

Year	Site	Sampling	Within perimeter strip		MC 10m distant to		MC 30m distant to		Growth stage	
		date	TR-PS	OSR-PS	TR-PS	OSR-PS	TR-PS	OSR-PS	TR	OSR
2009	HE	04/08	**1.3** (±0.2) a	**1.1** (±0.3) a	**1.0** (±0.2) a	**0.8** (±0.2) a	**1.3** (±0.1) a	**1.3** (±0.1) a	53	52
		04/11	**3.4** (±0.3) a	**2.7** (±0.6) a	**2.4** (±0.2) a	**2.1** (±0.4) a	**1.9** (±0.2) a	**2.2** (±0.2) a	55	55
		04/14	**3.9** (±0.1) a	**3.2** (±0.2) b	**2.7** (±0.2) b	**2.7** (±0.2) b	**2.4** (±0.1) b	**2.5** (±0.1) b	59	59
		04/19	**2.4** (±0.1) a	**1.1** (±0.2) b	**1.0** (±0.2) b	**1.1** (±0.1) b	**1.1** (±0.2) b	**1.0** (±0.1) b	61	61
		04/22	**1.5** (±0.2) a	**1.6** (±0.4) a	**1.8** (±0.5) a	**1.9** (±0.3) a	**1.7** (±0.3) a	**2.5** (±0.6) a	63	63
		04/25	**1.2** (±0.2) a	**1.2** (±0.2) a	**2.0** (±0.7) a	**1.2** (±0.3) a	1.0 (±0.3) a	**1.9** (±0.2) a	65	65
	LS	04/09	**2.0** (±0.3) a	**0.7** (±0.1) b	**0.5** (±0.1) b	**0.6** (±0.1) b	**0.7** (±0.1) b	**0.9** (±0.1) b	55	52
		04/12	**5.8** (±0.7) a	**2.6** (±0.4) b	**2.6** (±0.4) b	**2.3** (±0.4) b	**2.0** (±0.0) b	**2.5** (±0.1) b	60	57
		04/15	**4.6** (±0.6) a	**3.0** (±0.4) ab	**3.0** (±0.1) ab	**2.5** (±0.4) b	**3.1** (±0.1) ab	**2.7** (±0.3) ab	60	57
		04/20	**2.7** (±0.6) a	**3.9** (±0.3) a	**3.9** (±0.1) a	**3.7** (±0.1) a	**3.6** (±0.4) a	**3.1** (±0.2) a	61	59
		04/23	**0.9** (±0.2) a	**2.0** (±0.2) ab	**2.9** (±0.1) b	**2.8** (±0.2) b	**2.9** (±0.6) b	**2.2** (±0.3) ab	65	62
		04/26	**0.8** (±0.2) a	**2.2** (±0.2) ab	**3.1** (±0.4) b	**3.1** (±0.3) ab	**3.8** (±0.6) b	**3.3** (±0.6) b	65	65
2010	HE	04/13	**1.8** (±0.2) a	**0.5** (±0.1) b	**0.5** (±0.1) b	**0.5** (±0.1) b	**0.4** (±0.1) b	**0.5** (±0.1) b	57	51
		04/17	**4.9** (±0.5) a	**1.4** (±0.3) b	**1.4** (±0.1) b	**1.5** (±0.3) b	**1.1** (±0.2) b	**1.3** (±0.1) b	57	51
		04/20	**13.0** (±1.8) a	**4.9** (±1.3) b	**4.0** (±0.7) b	**5.1** (±0.6) b	**3.6** (±0.4) b	**4.1** (±0.2) b	59	52
		04/23	**8.4** (±0.7) a	**2.8** (±0.6) b	**2.5** (±0.5) b	**3.1** (±0.3) b	**2.1** (±0.2) b	**2.8** (±0.2) b	59	53
		04/26	**16.8** (±2.2) a	**8.4** (±2.2) b	**7.5** (±1.5) b	**9.2** (±1.4) b	**6.8** (±0.2) b	**7.6** (±1.1) b	60	56
		04/29	**16.6** (±0.4) a	**7.5** (±1.3) b	**7.5** (±0.5) b	**7.7** (±0.5) b	**6.4** (±0.6) b	**6.1** (±0.6) b	63	57
		05/02	**5.7** (±0.7) a	**5.3** (±1.2) a	**4.9** (±0.9) a	**5.0** (±0.7) a	**4.8** (±0.7) a	**4.3** (±0.1) a	65	60
		05/05	**3.1** (±0.8) a	**3.0** (±0.6) a	**3.4** (±0.4) a	**2.8** (±0.2) a	**2.6** (±0.5) a	**2.5** (±0.1) a	67	61
		05/09	**2.4** (±0.7) a	**3.0** (±0.5) a	**3.2** (±0.3) a	**2.9** (±0.1) a	**2.5** (±0.3) a	**3.2** (±0.1) a	67	62

Discussion

From our experiments it was obvious that cabbage stem flea beetle and pollen beetle preferred the turnip rape trap crop to oilseed rape. This was also found by other authors in studies on the host plant preference of insect pests for various Brassica species (Büchi, 1995; Barari et al., 2005; Büchs, 2009). In contrast, the three species of *Ceutorhynchus* recorded in our experiments did not show a preference for turnip rape. In former studies, contradictory results on the infestation level of oilseed rape and turnip rape by *Ceutorhynchus* spp. were found (Büchi, 1990; Kalischuk & Dosdall, 2004). Cabbage seed weevil preferred oilseed rape for oviposition in experiments conducted in Switzerland, while turnip rape was preferred to oilseed rape in experiments performed in Canada (Büchi, 1990; Kalischuk & Dosdall, 2004). Cabbage stem weevil preferred turnip rape to oilseed rape in experiments in the UK, while the abundance of this species on turnip rape was not higher than on oilseed rape in Switzerland (Büchi, 1990; 1995; Barari et al., 2005). In our field trials the preference of pollen beetle for turnip rape was regularly observed during the early immigration phase. The difference between beetle numbers on turnip rape and oilseed rape was more obvious when turnip rape development was more advanced than oilseed rape. Host selection of pollen beetle is strongly influenced by the phenology of the host plant (Hokkanen et al., 1986; Frearson et al., 2005; Cook et al., 2007). Adult beetles preferred plants in an advanced growth stage to backward plants (Hokkanen et al., 1986; Nilsson, 2004; Frearson et al., 2005; Cook et al., 2006; Cook et al., 2007). Moreover, pollen beetles prefer turnip rape over oilseed rape due to volatile emissions of turnip rape (Cook et al., 2006; Cook et al., 2007). Many other specialist insect pests on Brassicaceae are known to locate their host plant via olfactory stimuli (Bartlet, 1996; Walczak et al., 1998). One major group of attractive volatiles are isothiocyanates (Bartlet et al., 1993; Bartlet et al., 1997). These volatiles are produced at degradation of glucosinolates in damaged plant cells (Bartlet et al., 1993). Increased amounts of isothiocyanates are released following destruction of cells due to herbivore feeding (Rask et al., 2000). Infestation by insects is known to induce glucosinolate accumulation in the plant tissues (Chapter II) (Koritsas et al., 1989; 1991; Birch et al., 1992; Bartlet et al., 1999). Therefore, the higher infestation of turnip rape by cabbage stem flea beetle may have been responsible for an increased attraction of subsequent herbivores like pollen beetle. Plant morphology such as plant architecture could provide other important factors for host selection, as described by the "plant vigor hypothesis" (Price, 1991). Based on literature review, Price (1991) stated that many insect species prefer vigorous host plant over backward plants. Previous studies support this hypothesis also for the cabbage stem flea beetle which preferred plants of early sown oilseed rape of late sown oilseed rape (Thioulouse, 1987). Nuss (2004) found a higher density of cabbage stem flea beetle larvae at high sowing rates of oilseed rape as compared to low sowing rates. This author supposed that the higher attractiveness of higher plant densities is derived from the denser crop canopy. The effect of plant size on host selection was also shown for many other herbivorous insect species (Cornelissen et al., 2008). Therefore the role of plant architecture, plant density and leaf area index for efficacy of trap crops needs to be investigated in further experiments in greater detail.

For none of the insect species studied in our experiments the establishment of turnip rape perimeter strips was associated with reduced levels of infestation on the adjacent main crop. Accordingly, in these plots the plant damage was not reduced and the seed yield was not higher compared to oilseed rape plots adjacent to oilseed rape perimeter strips. Previous studies dealing with the efficacy of turnip rape perimeter strips have shown inconsistent results, depending on the pest species investigated. Only little or no effect of turnip rape perimeter strips on the infestation by *Ceutorhynchus* species of the main crop could be detected (Büchi, 1995; Cook et al., 2004; Barari et al., 2005). In contrast, for cabbage stem flea beetle and pollen beetle a significant reduction of the infestation level of the main crop was reported in the UK (Cook et al., 2004; Barari et al., 2005). Similar results were obtained from pollen beetle when an early flowering oilseed rape cultivar was grown as trap crop in the perimeter strip (Hokkanen et al., 1986). In many previous studies spring cultivars of oilseed rape and turnip rape were grown (Hokkanen et al., 1986; Cook et al., 2004). This may have had an effect on the efficacy of the trap cropping system, because spring crops are often attacked by new generation beetles which may show different behaviour than beetles after winter diapause. Other studies showed that the efficacy of the trap crop can be increased by targeted spraying of insecticides, even for the cabbage seed weevil

(Büchi, 1995; Buntin, 1998; Carcamo et al., 2007). On the contrary, Barari et al. (2005) found little effect of insecticide application targeted at the perimeter strip on the infestation by cabbage stem flea beetles and cabbage stem weevils on the main crop.

In our experiments immigrating pollen beetles first colonized the turnip rape in the perimeter strip. The oilseed rape plants in the centre of the field were not attacked until they reached the end of the bud stage, a highly attractive growth stage for pollen beetles (Fearson et al. 2005). Moser et al. (2009) showed that pollen beetles are very mobile and are able to locate their host plants over a long distance. For this reason a perimeter strip of highly attractive turnip rape may attract pollen beetles over a long distance, which may increases the risk of even higher levels of infestation of the main crop compared to a crop without turnip rape perimeter strips. However, our results did not support this hypothesis and earlier studies also did not report higher infestation levels by pests resulting from turnip rape trap cropping (Nilsson, 2004).

In previous studies, turnip rape cultivars containing high amounts of glucosinolates in the seeds were commonly used (Büchi, 1990; 1995). To avoid negative effects on the quality of the harvested oilseed rape seed, plots grown with these cultivars have to be destroyed before harvest. In our experiments, the turnip rape cultivar 'Largo' showing double low quality of the seed (low contents of erucic acid and glucosinolates) was selected. By using this cultivar, the trap crop would not only reduce pest densities on the main crop, but also could be harvested together with the main crop, thereby avoiding yield losses due an unproductive area grown by a trap crop with a high glucosinolate content in the seeds. However, the turnip rape cultivar 'Largo' was not appropriate for reducing the infestation levels by pests in the oilseed rape main crop. Other turnip rape cultivars with double low quality in the seeds have shown low potential to increase the efficacy of this trap cropping system (Chapter IV).

In conclusion, in our experiments growing of turnip rape perimeter strips as trap crops provided no suitable strategy for management of oilseed rape pests. In future field trials, the efficacy of trap cropping combined with insecticide application targeted to the perimeter strip needs to be evaluated (Büchi 1995; Barari et al. 2005). Furthermore, a careful selection of a highly attractive trap crop and a less attractive main crop cultivar may increase the efficacy of this system (Cook et al., 2003; Williams, 2004). Oilseed rape lines were reported recently from Canada which show resistance to insect feeding and are less attractive than conventional oilseed rape cultivars (Tansey et al., 2011). A combination of these resistant oilseed rape cultivars and a highly attractive trap crop have potential to be applied as “push – pull strategy” against insect pests (Cook et al., 2003; Cook et al. 2007).

References

Alford DV (2003) The oilseed rape crop. In: Biocontrol of oilseed rape pests (ed. DV Alford). Blackwell Science Ltd., Oxford, pp. 1-8.

Alford DV, Nilsson C & Ulber B (2003) Insect pests of oilseed rape crops. In: Biocontrol of oilseed rape pests (ed. DV Alford). Blackwell Science Ltd., Oxford, pp. 9-42.

Attah PK & Lawton JH (1984) The invasion of rape crops by insect pests, with special reference to pollen beetle, *Meligethes* spp. Aspects of Applied Biology 6: 265-274.

Barari H, Cook SM, Clark SJ & Williams IH (2005) Effect of a turnip rape (*Brassica rapa*) trap crop on stem-mining pests and their parasitoids in winter oilseed rape (*Brassica napus*). Biocontrol 50: 69-86.

Bartlet E (1996) Chemical cues to host-plant selection by insect pests of oilseed rape. Agricultural Zoology Reviews 7: 89-116.

Bartlet E, Blight MM, Hick AJ & Williams IH (1993) The responses of the cabbage seed weevil (*Ceutorhynchus assimilis*) to the odour of oilseed rape (*Brassica napus*) and to some volatile isothiocyanates. Entomologia Experimentalis et Applicata 68: 295-302.

Bartlet E, Blight MM, Lane P & Williams IH (1997) The responses of the cabbage seed weevil *Ceutorhynchus assimilis* to volatile compounds from oilseed rape in a linear track olfactometer. Entomologia Experimentalis et Applicata 85: 257-262.

Bartlet E, Kiddle G, Williams I & Wallsgrove R (1999) Wound-induced increases in the glucosinolate content of oilseed rape and their effect on subsequent herbivory by a crucifer specialist. Entomologia Experimentalis et Applicata 91: 163-167.

Birch ANE, Griffiths DW, Hopkins RJ, Smith WHM & McKinlay RG (1992) Glucosinolate responses of swede, kale, forage and oilseed rape to root damage by turnip root fly (*Delia floralis*) larvae. Journal of the Science of Food and Agriculture 60: 1-9.

Bommarco R, Miranda F, Bylund H & Bjorkman C (2011) Insecticides suppress natural enemies and increase pest damage in cabbage. Journal of Economic Entomology 104: 782-791.

Büchi R (1990) Investigation on the use of turnip rape as a trap plant to control oilseed rape pests. IOBC/wprs Bulletin 13 (4): 32-39.

Büchi R (1995) Combination of trap plants (*Brassica rapa var. vestris*) and insecticide use to control rape pests. IOBC/wprs Bulletin 18 (4): 102-119.

Büchs W (2009) Fangpflanzen (trap crop) als Methode zur Kontrolle tierischer Schädlinge im ökologischen Rapsanbau. Mitteilungen der Deutschen Gesellschaft für Allgemeine und Angewandte Entomologie 17: 241-246.

Buntin GD (1998) Cabbage seedpod weevil (*Ceutorhynchus assimilis*, Paykull) management by trap cropping and its effect on parasitism by *Trichomalus perfectus* (Walker) in oilseed rape. Crop Protection 17: 299-305.

Carcamo HA, Dunn R, Dosdall LM & Olfert O (2007) Managing cabbage seedpod weevil in canola using a trap crop - A commercial field scale study in western Canada. Crop Protection 26: 1325-1334.

Cook SM, Rasmussen HB, Birkett MA, Murray DA, Pye BJ, Watts NP & Williams IH (2007) Behavioural and chemical ecology underlying the success of turnip rape (*Brassica rapa*) trap crops in protecting oilseed rape (*Brassica napus*) from the pollen beetle (*Meligethes aeneus*). Arthropod-Plant Interactions 1: 57-67.

Cook SM, Khan ZR & Pickett JA (2007) The use of push-pull strategies in integrated pest management. Annual Review of Entomology 52: 375-400.

Cook SM, Smart LE, Martin JL, Murray DA, Watts NP & Williams IH (2006) Exploitation of host plant preferences in pest management strategies for oilseed rape (Brassica napus). Entomologia Experimentalis et Applicata 119: 221-229.

Cook SM, Smart LE, Potting RJP, Bartlet E, Martin JL, Murray DA, Watts NP & Williams I (2002) Turnip rape (*Brassica rapa*) as a trap crop to protect oilseed rape (*Brassica napus*) from infestation by insect pests: potential and mechanisms of action. The BCPC Conference: Pests & Diseases 2002 2: 569-574.

Cook SM, Smart LE, Rasmussen HB, Bartlet E, Martin JL, Murray DA, Watts NP & Williams IH (2003) Push-Pull strategies to reduce insecticide input to oilseed rape (*Brassica napus*): Potential of low alkenyl glucosinolate oilseed rape varieties (push!), and turnip rape (*Brassica rapa*) trap crops (pull!): 11th International Rapeseed Congress Copenhagen, pp. 1018-1021.

Cook SM, Watts NP, Hunter F, Smart LE & Williams IH (2004) Effects of a turnip rape trap crop on the spatial distribution of *Meligethes aeneus* and *Ceutorhynchus assimilis* in oilseed rape. IOBC/wprs Bulletin 27 (10): 199-206.

Cornelissen T, Fernandes GW & Vasconcellos-Neto J (2008) Size does matter: variation in herbivory between and within plants and the plant vigor hypothesis. Oikos 117: 1121-1130.

Eickermann M, Ulber B & Vidal S (2010) Resynthesized lines and cultivars of *Brassica napus* L. provide sources of resistance to the cabbage stem weevil (*Ceutorhynchus pallidactylus* (Marsh.)). Bulletin of Entomological Research 110: 1-8.

Ferguson AW, Klukowski Z, Walczak B, Clark SJ, Mugglestone MA, Perry JN & Williams IH (2003) Spatial distribution of pest insects in oilseed rape: implications for integrated pest management. Agriculture Ecosystems & Environment 95: 509-521.

Frearson DJT, Ferguson AW, Campbell JM & Williams IH (2005) The spatial dynamics of pollen beetles in relation to inflorescence growth stage of oilseed rape: implications for trap crop strategies. Entomologia Experimentalis et Applicata 116: 21-29.

Fritzsche R (1957) Zur Biologie und Ökologie der Rapsschädlinge aus der Gattung *Meligethes*. Journal of Applied Entomology 40: 222-280.

Giamoustaris A & Mithen R (1996) The effect of flower colour and glucosinolates on the interaction between oilseed rape and pollen beetles. Entomologia Experimentalis et Applicata 80: 206-208.

Heimbach U & Müller A (2011) Pyrethroid resistance of oilseed rape pest insects from 2005-2010 and German insecticide resistance management strategy. 13th International Rapeseed Congress Prague, pp. 1278-1281.

Hokkanen H, Granlund H, Husberg GB & Markkula M (1986) Trap crops used successfully to control *Melighethes aeneus* (Col, Nitidulidae), the rape blossom beetle. Annales Entomologici Fennici 52: 115-120.

Hokkanen HMT (1991) Trap cropping in pest management. Annual Review of Entomology 36: 119-138.

Kalischuk AR & Dosdall LM (2004) Susceptibilities of seven Brassicaceae species to infestation by the cabbage seedpod weevil (Coleoptera : Curculionidae). Canadian Entomologist 136: 265-276.

Koritsas VM, Lewis JA & Fenwick GR (1989) Accumulation of indole glucosinolates in *Psylliodes chrysocephala* L. infested, or *Psylliodes chrysocephala* L. damaged tissues of oilseed rape (*Brassica napus* L.). Experientia 45: 493-495.

Koritsas VM, Lewis JA & Fenwick GR (1991) Glucosinolate response of oilseed rape, mustard and kale to mechanical wounding and infestation by cabbage stem beetle (*Psylliodes chrysocephala*). Annals of Applied Biology 118: 209-221.

Lambdon PW, Hassall M & Mithen R (1998) Feeding preferences of woodpigeons and flea-beetles for oilseed rape and turnip rape. Annals of Applied Biology 133: 313-328.

Lancashire PD, Bleiholder H, Vandenboom T, Langeluddeke P, Stauss R, Weber E & Witzenberger A (1991) A uniform decimal code for growth-stages of crops and weeds. Annals of Applied Biology 119: 561-601.

Lerin J (1993) Influence of the growth rate of oilseed rape on the splitting of the stem after an attack of *Ceutorhynchus napi* Gyll. IOBC/wprs Bulletin 16 (3): 160-163.

Moser D, Drapela T, Zaller JG & Frank T (2009) Interacting effects of wind direction and resource distribution on insect pest densities. Basic and Applied Ecology 10: 208-215.

Nilsson C (2004) Trap plants to avoid insecticide application against pollen beetles in oilseed rape. IOBC/wprs Bulletin 27 (10): 215-221.

Nuss H (2004) Einfluss der Pflanzendichte und -architektur auf Abundanz und innerpflanzliche Verteilung stängelminierender Schadinsekten in Winterraps. PhD-thesis, Georg-Universität-Universität Göttingen.

Price PW (1991) The plant vigor hypothesis and herbivore attack. Oikos 62: 244-251.

Rask L, Andreasson E, Ekbom B, Eriksson S, Pontoppidan B & Meijer J (2000) Myrosinase: gene family evolution and herbivore defense in Brassicaceae. Plant Molecular Biology 42: 93-113.

Slater R, Ellis S, Genay JP, Heimbach U, Huart G, Sarazin M, Longhurst C, Müller A, Nauen R, Rison JL & Robin F (2011) Pyrethroid resistance monitoring in European populations of pollen beetle (*Meligethes* spp.): a coordinated approach through the Insecticide Resistance Action Committee (IRAC). Pest Management Science 67: 633-638.

Tansey JA, Dosdall LM & Keddie A (2011) Incorporation of novel *Ceutorhynchus obstrictus*-resistant canola genotypes into mixed cropping strategies: evidence for associational resistance. Journal of Insect Behavior 24: 116-131.

Thioulouse J (1987) Space-time structures in a winter rape pest population, *Psylliodes chrysocephala* (Col. Chrysomelidae) methodological proposals and biological interpretations. Journal of Applied Ecology 24: 435-450.

Walczak B, Kelm M, Klukowski Z, Smart L, Ferguson A & Williams I (1998) The effect of trap design and 2-phenylethyl isothiocyanate on catches of stem weevils (*Ceutorhynchus pallidactylus* Marsh. and *C. napi* Gyll.) in winter oilseed rape. IOBC/wprs Bulletin 21 (5): 141-146.

Williams IH (2004) Advances in insect pest management of oilseed rape in Europe. In: Insect pest management (ed. AR Horowitz & I Ishaaya). Springer-Verlag, Berlin Heidelberg, pp. 181-208.

Williams IH (2010) The major insect pests of oilseed rape in europe and their management: an overview. In: Biocontrol-based integrated management of oilseed rape pests (ed. IH Williams). Springer Science+Business B.V., Dordrecht, Heidelberg, London, New York, pp. 1-43.

Williams IH & Free JB (1978) The feeding and mating behaviour of pollen beetles (*Meligethes aeneus* Fab.) and seed weevils (*Ceutorhynchus assimilis* Payk.) on oil-seed rape (*Brassica napus* L.). Journal of Agricultural Science 91: 453-459.

Chapter VI: General discussion

Screening of cultivars

In the past, only a few studies have investigated the adult feeding preference of cabbage stem flea beetle (Lambdon et al., 1998; Williams & Bartlet, 1993). The performance of larvae on different host plant cultivars and species has been rarely assessed before (Godan, 1951). This knowledge is essential for the development of integrated pest management strategies, like trap cropping, which aims at manipulating the insect spatial distribution on the crop (Cook et al., 2006b). The results of our experiments showed a significantly better performance of larvae grown on turnip rape than of larvae grown on oilseed rape. In comparison to these plant species, the performance of larvae on white mustard (*Sinapis alba*) was significantly reduced. Similar effects of white mustard and oilseed rape were also observed with larvae of cabbage seed weevil in Canada (Kalischuk & Dosdall, 2004; Tansey et al., 2010). The mechanisms underlying the larval performance on different host plants have not been identified. Glucosinolates as the predominant class of secondary plant metabolites in Brassicaceae were supposed to play a major role for the development of larvae (Agrawal & Kurashige, 2003; Uddin et al., 2008). Therefore, many researchers have focused on the impact of these chemicals to larval performance of various specialist insects on Brassicaceae (Agrawal & Kurashige, 2003; Ulmer & Dosdall, 2006; Uddin et al., 2008; Uddin et al., 2009). The results obtained from these studies are inconsistent and sometimes contradictory. Mewis et al. (2008) did not find an effect of modified aliphatic and indole glucosinolate profiles on performance of the larvae of *Pieris rapae* and *P. brassicae*. Tansey et al. (2010) observed a positive correlation between larval growth parameters of cabbage seed weevil and gluconasturtiin, while larval growth was negatively correlated to neoglucobrassicin. In our experiments, a positive correlation between the larval performance of cabbage stem flea beetle and the content of progoitrin and 4-hydroxyglucobrassicin in the tested petioles was shown. Further, plant responses induced by larval feeding, especially the change of glucosinolate composition, in our tests differed substantially from earlier studies reported in the literature (Koritsas et al., 1989; 1991; Bartlet et al., 1999).

Following larval feeding, we observed an induced increase of aliphatic glucosinolates in newly grown, not damaged leaves of most brassicaceous species tested while in other studies an increase of indole glucosinolates in infested oilseed rape plants has been reported (Koritsas et al., 1989; 1991; Bartlet et al., 1999).

The results reported on the effects of glucosinolates on adult feeding preference are contradictory as well. Many researchers concluded that glucosinolates act as feeding stimulants for specialist herbivorous insects (Bartlet et al., 1994; Schoonhoven et al., 2005). However, the effect of single glucosinolate compounds was very variable. Giamoustaris & Mithen (1995) for example found a higher potential of gluconapin than of progoitrin and glucobrassicanapin for feeding stimulation of adult cabbage stem flea beetles. Previous studies have shown inconsistent and sometimes contradictory results, depending on the insect species used in the experiments (Bodnaryk & Palaniswamy, 1990; Siemens & MitchellOlds, 1996; Bartlet et al., 1996; Rohr et al., 2009; Tansey & Dosdall, 2011). Therefore, it seems unlikely that our results can be conveyed directly to other insect pests on oilseed rape, for example the rape stem weevil or cabbage stem weevil. Glucosinolates are the most important group of secondary plant substances in Brassicaceae (Mithen, 2001), but there are some indications that they are not the only plant cues affecting the feeding behaviour and larval performance of cabbage stem flea beetle (Bartlet & Williams, 1991; Bartlet et al., 1994). In this study, additional analyses were conducted to identify other plant characteristics which also may play an important role for host preference of insects: neither sugar composition nor phenolics in the leaf lamina were correlated to the feeding behaviour of cabbage stem flea beetle. In contrast, earlier studies have shown significant effects of these substances on feeding behaviour of insects, acting either as feeding stimulants or as deterrents (Bernays, 1981; Meisner & Mitchell, 1984; Bartlet et al., 1994; Schoonhoven et al., 2005). A high trichome density of the leaf surface as well did not reduce adult feeding of cabbage stem flea beetle compared to leaves with a low trichome density. That might be explained by the structure of the trichomes. Lamb (1980) reported that soft trichomes of turnip rape do not deter feeding of cabbage flea beetle, contrasting to the stiff trichomes of *Brassica hitra* which reduced cabbage flea beetle feeding significantly. In future experiments, a wider range of host plants needs to be included into the experiments, which may provide a higher variability of specific plant cues. There is a huge number of potential host plants for cabbage stem flea beetle that have not been tested for their host plant quality (Godan, 1951; Williams & Bartlet, 1993). Alternatively, an artificial diet has been developed which contains different amounts of sugars and other plant metabolites that are supposed to play a role for the feeding preference and performance of larvae (Bartlet et al., 1994). Different leaf substances such as sugars and glucosinolates may act synergistically (Bartlet et al., 1994). These effects may have masked the effect of single substances in our experiments.

The adult feeding preference of cabbage stem flea bettle can differ from oviposition preference. Other insect pests on oilseed rape, such as pollen beetle or seed weevil, are known to be more selective in accepting a host plant for oviposition than for feeding (Dmoch, 1965; Free & Williams, 1978).

For many herbivorous insect species a positive correlation between adult oviposition preference and larval performance was hypothesized (Pöykkö, 2006; Shikano et al., 2010). In the present experiments a significant correlation of adult feeding preference and larval performance of cabbage stem flea beetle was observed on the tested cultivars of oilseed rape and turnip rape. However, this correlatio was not proven on white mustard. Therefore, in future experiments the "preference-performance hypothesis" needs to be evaluated for cabbage stem flea beetle in greater detail. Further, a screening method appropriate for testing the oviposition preference of cabbage stem flea beetle females under laboratory conditions needs to be developed. This method could be used to determine whether adult feeding preference and oviposition is correlated, especially in white mustard.

The results obtained from our field plot experiments were similar to the results obtained in the laboratory. Under both conditions the cabbage stem flea beetle significantly preferred turnip rape to oilseed rape, while there was no significant difference in host preference between turnip rape cultivars. Results obtained in the field provide only an indirect estimation of adult oviposition preference, as the number of larvae in the plants is not only determined by the number of eggs deposited per plant by females. Many other factors such as egg and larval mortality due to predators or to insufficient soil moisture were not taken into account (Kaufmann, 1941; Schulz, 1985, Williams et al., 2010). Moreover, the impact of host plant species and cultivar on larval performance could not be measured precisely in the field plot experiments. In our laboratory experiments, the larval weight differed significantly between host plant species.

Trap cropping

In the screening of cultivars under field conditions, turnip rape was found to be highly attractive for pollen beetle and cabbage stem flea beetle. These results confirm earlier studies which also showed a high attractiveness of turnip rape for these insect species (Büchi, 1990; Lambdon et al., 1998; Barari et al., 2005; Cook et al., 2006b). In contrast, plant infestation by rape stem weevil and cabbage stem weevil was not affected by plant species or cultivar. The cabbage seed weevil even preferred oilseed rape to turnip rape for oviposition. In these field experiments, only the pollen beetle discriminated between different turnip rape cultivars. Presumably the growth stage of the host plant is the major factor for this response (Williams & Free, 1978; Nilsson, 2004; Frearson et al., 2005). Prior studies showed that other specialised oilseed rape pests, such as cabbage stem weevil or cabbage stem flea beetle, discriminate between different oilseed rape genotypes (Williams, 1989; Lambdon et al., 1998; Eickermann et al., 2010). These results indicate that a wider range of plant species should be included into future experiments, in order to provide a higher variability in the attractiveness to pest species. It is also necessary to identify the plant cues that determine the host location to herbivorous insects. Plant volatiles may play an important role as they are known to attract insects over a long distance (Schoonhoven et al., 2005). Visual cues derived from plants or non volatile chemical stimuli and leaf morphology can also determine the plant attractiveness and acceptance to pests (Giamoustaris & Mithen, 1996; Schoonhoven et al., 2005; Cook et al., 2006a). Furthermore, experiments are needed to identify the influence of plant architecture on host selection of cabbage stem flea beetle as previous studies showed an effect of these factors on oilseed rape pests (Dosdall et al., 1996; Nuss, 2004). This knowledge may help to identify additional plant factors which determine the attractiveness of a potential trap crop. Overall, in our experiments under laboratory and field conditions turnip rape proved to be a suitable plant species for trap cropping.

When turnip rape was tested as a trap crop, either sown as a perimeter strip or sown in mixtures with oilseed rape overall, it failed to reduce the infestation of the oilseed rape main crop by pest insects and the damage caused by these insects on plants. Neither the turnip rape sown as perimeter strip around an oilseed rape field nor a mixture of both plant species resulted in a significant yield increase of the main crop. In the past, much work has been done on intercropping of two plant species and in many experiments this system significantly reduced pest infestation of the main crop (Tukahirwa & Coaker, 1982; Theunissen et al., 1995; Nilsson, 2004). Contrasting to our experiments, in previous intercropping experiments of brassicaceous crops mostly a host plant and a non host plant were sown in the mixture, as has been reviewed by Finch & Collier (2000). The effect of intercrops on pest infestation has been explained by different hypotheses (see Altieri, 1994). Two very popular hypotheses are the "Resource Concentration Hypothesis" and the "Enemies Hypothesis" which have been first presented by Root (1973). In the first hypothesis the reduction of pest infestation is explained by the preference of herbivorous insects for host plants when these are aggregated. In intercrops the host plants are less concentrated and therefore less attractive to pest species (Root, 1973). The 'enemy hypothesis' explains the lower pest infestation in intercrops by the encouragement of natural enemies of pest species due to a more diverse environment (Root, 1973).

Finch & Collier (2000) stated that host location and acceptance of herbivorous insects is a chain of three steps. The first two steps are host location by using olfactory and visual cues which initiate landings on the plant surface (Finch & Collier, 2000). On the plant the insects receive stimuli from non volatile leaf substances (Finch & Collier, 2000). Only if the plant has provided appropriate cues, it will be accepted as a host (Finch & Collier, 2000). In monocultures all landings will be on a host plant and are most often followed by final host acceptance (Finch & Collier, 2000). In intercrops where host plants and non host plants are growing close to each other, landings on non host plants will interrupt the behavioural host finding chain (Finch & Collier, 2000). This hypothesis might explain the low efficacy of intercropping by using two host plant species, as has been conducted in our experiments. On the other hand, some experiments in the past have shown promising results for the management of pollen beetles by intercropping of oilseed rape and turnip rape (Büchi, 1990; Nilsson, 2004; Strauch, 2010).

Earlier studies have shown the potential of turnip rape perimeter strips for control of insect pests on oilseed rape (Büchi, 1995; Cook et al., 2004; Carcamo et al., 2007a). However, we did not find evidence for the suitability of this trap cropping system. Various factors might be responsible for the efficacy of turnip rape trap crops. One of these is the size of the perimeter strips. Hannunnen (2005) underlined that the proportion of the trap crop area relative to the main crop has impact on the efficacy of this system. Thus, one possible reason for the failure of intercropping of oilseed rape and turnip rape and of turnip rape perimeter strips in our study might be the low proportion of trap crop area relative to the main crop area. Furthermore, as shown in our field plot experiments and also mentioned in the literature, the turnip rape trap crop is not attractive to species of *Ceutorhynchus* (Büchi, 1990; 1995; Cook et al., 2006b). Therefore, turnip rape is not a suitable trap crop for stem weevils and cabbage seed weevil.

In the literature it was argued that trap cropping by using a highly attractive plant species might attract even higher numbers of pests to the field resulting in higher infestation levels compared to a field without turnip rape (Nilsson, 2004). However, our results did not support this hypothesis, neither in mixed cropping nor in perimeter strip systems.

Conclusions

The efficacy of trap cropping may be improved by a targeted selection of the trap and main crop (Cook et al., 2003). This may be achieved by increasing the differences in relative attractiveness between the two crop species for pests (Cook et al., 2003). Results from field plot experiments showed that 'Malwira' was the most attractive turnip rape cultivar for pollen beetle. It was as attractive as 'Largo' and 'Perko' for other pest species. Therefore 'Malwira' may have potential to increase the efficacy of both trap cropping systems. However, the high content of erucic acid and glucosinolates in the seeds may reduce the suitability of this system for practical application (Bell, 1995; Uppström, 1995).

So far only turnip rape is known to be very attractive to many specialised insect pests on oilseed rape. A screening of a wider set of plant species may identify trap crops which are more attractive to *Ceutorhynchus* species than turnip rape. Further, a "multi trap cropping system", as described by Shelton & Badenes-Perez (2006), of two or more plant species may affect a wider range of insect species over a longer period.

Previous studies showed that the parasitism rate of cabbage stem weevil larvae and pollen beetle larvae was higher in turnip rape than in oilseed rape (Linz, 1991, Barari et al., 2005). Therefore, it needs to be checked whether a turnip rape perimeter strip may also affect the parasitism rate of pest larvae in the main crop. This could help to reduce the population density on long term and improve the integrated pest management strategy (Ulber et al., 2010).

Insecticide applications targeted to the trap crop perimeter strips could also enhance the efficacy of turnip rape perimeter strips and would limit the application of insecticides to a small area of the oilseed rape field (Williams, 2004). Some prior studies showed promising results, while this system failed in other experiments (Büchi, 1995; Buntin, 1998; Barari et al., 2005; Carcamo et al., 2007a). One reason for the low efficacy in some studies might be a suboptimal timing of insecticide application. Therefore, this system needs to be assessed further in future experiments.

The breeding of cultivars resistant to pest insects seems to be promising as well in order to overcome actual insect pest challenges. Recently sources of resistance have been identified in some brassicaceous crops (Cole, 1994; Carcamo et al., 2007b; Cartea et al., 2010; Eickermann et al., 2010). Moreover, the combination of growing resistant cultivars and trap cropping might be used as a 'push' and 'pull' strategy (Cook et al., 2007).

References

Agrawal AA & Kurashige NS (2003) A role for isothiocyanates in plant resistance against the specialist herbivore *Pieris rapae*. Journal of Chemical Ecology 29: 1403-1415.

Altierie MA (1994) Biodiversity and pest management in agroecosystems. Haworth Press Inc.,New York: 185pp. Cited in: Finch S & Collier RH (2000) Host-plant selection by insects - a theory based on 'appropriate/inappropriate landings' by pest insects of cruciferous plants. Entomologia Experimentalis et Applicata 96: 91-102.

Barari H, Cook SM, Clark SJ & Williams IH (2005) Effect of a turnip rape (*Brassica rapa*) trap crop on stem-mining pests and their parasitoids in winter oilseed rape (*Brassica napus*). Biocontrol 50: 69-86.

Bartlet E, Kiddle G, Williams I & Wallsgrove R (1999) Wound-induced increases in the glucosinolate content of oilseed rape and their effect on subsequent herbivory by a crucifer specialist. Entomologia Experimentalis et Applicata 91:163-167.

Bartlet E, Mithen R & Clark SJ (1996) Feeding of the cabbage stem flea beetle *Psylliodes chrysocephala* on high and low glucosinolate cultivars of oilseed rape. Entomologia Experimentalis et Applicata 80: 87-89.

Bartlet E, Parsons D, Williams IH & Clark SJ (1994) The influence of glucosinolates and sugars on feeding by the cababge stem flea beetle, *Psylliodes chrysocephala*. Entomologia Experimentalis et Applicata 73: 77-83.

Bartlet E & Williams IH (1991) Factors restricting the feeding of the cabbage stem flea beetle (*Psylliodes chrysocephala*). Entomologia Experimentalis et Applicata 60: 233-238.

Bell JM (1995) Meal and by-product utilization in animal nutrition. In: Brassica oilseeds - production and utilization (ed. DS Kimber & DI McGregor). CAB International, Wallingford, pp. 301-338.

Bernays EA (1981) Plant tannins and insect herbivores - an appraisal. Ecological Entomology 6: 353-360.

Bodnaryk RP & Palaniswamy P (1990) Glucosinolate levels in cotyledons of mustard, *Brassica juncea* L. and rape, *Brassica napus* L. do not determine feeding of flea beetle *Phyllotreta cruciferae* (Goeze). Journal of Chemical Ecology 16: 2735-2746.

Büchi R (1990) Investigation on the use of turnip rape as a trap plant to control oilseed rape pests. IOBC/wprs Bulletin 13 (4): 32-39.

Büchi R (1995) Combination of trap plants (*Brassica rapa var. vestris*) and insecticide use to control rape pests. IOBC/wprs Bulletin 18 (4): 102-119.

Buntin GD (1998) Cabbage seedpod weevil (*Ceutorhynchus assimilis*, Paykull) management by trap cropping and its effect on parasitism by *Trichomalus perfectus* (Walker) in oilseed rape. Crop Protection 17: 299-305.

Carcamo HA, Dunn R, Dosdall LM & Olfert O (2007a) Managing cabbage seedpod weevil in canola using a trap crop - A commercial field scale study in western Canada. Crop Protection 26: 1325-1334.

Carcamo H, Olfert O, Dosdall L, Herle C, Beres B & Soroka J (2007b) Resistance to cabbage seedpod weevil among selected Brassicaceae germplasm. Canadian Entomologist 139: 658-669.

Cartea ME, Francisco M, Lema M, Soengas P & Velasco P (2010) Resistance of cabbage (*Brassica oleracea capitata* Group) crops to *Mamestra brassicae*. Journal of Economic Entomology 103: 1866-1874.

Cole RA (1994) Locating a resistance mechanism to the cabbage aphid in two wild brassicas. Entomologia Experimentalis et Applicata 71: 23-31.

Cook SM, Skellern MP, Smith MJ & Williams IH (2006a) Response of pollen beetles (*Meligethes aeneus*) to petal colour. IOBC/wprs Bulletin 29 (7): 151-158.

Cook SM, Smart LE, Martin JL, Murray DA, Watts NP & Williams IH (2006b) Exploitation of host plant preferences in pest management strategies for oilseed rape (*Brassica napus*). Entomologia Experimentalis et Applicata 119: 221-229.

Cook SM, Smart LE, Rasmussen HB, Bartlet E, Martin JL, Murray DA, Watts NP & Williams IH (2003) Push-Pull strategies to reduce insecticide input to oilseed rape (*Brassica napus*): Potential of low alkenyl glucosinolate oilseed rape varieties (push!), and turnip rape (*Brassica rapa*) trap crops (pull!): 11th International Rapeseed Congress Copenhagen, pp. 1018-1021.

Cook SM, Watts NP, Hunter F, Smart LE & Williams IH (2004) Effects of a turnip rape trap crop on the spatial distribution of *Meligethes aeneus* and *Ceutorhynchus assimilis* in oilseed rape. IOBC/wprs Bulletin 27 (10): 199-206.

Cook SM, Khan ZR & Pickett JA (2007) The use of push-pull strategies in integrated pest management. Annual Review of Entomology 52: 375-400.

Dmoch J (1965) The dynamics of population of the cabbage seedpod weevil (*Ceutorhynchus assimilis* Payk.) and the development of winter rape. Part 2. Ekologia Polska Seria A 8: 463-489.

Dosdall LM, Herbut MJ, Cowle NT & Micklich TM (1996) The effect of seeding date and plant density on infestations of root maggots, *Delia* spp, (Diptera: Anthomyiidae), in canola. Canadian Journal of Plant Science 76: 169-177.

Eickermann M, Ulber B & Vidal S (2010) Resynthesized lines and cultivars of *Brassica napus* L. provide sources of resistance to the cabbage stem weevil (*Ceutorhynchus pallidactylus* (Marsh.)). Bulletin of Entomological Research 110: 1-8.

Finch S & Collier RH (2000) Host-plant selection by insects - a theory based on 'appropriate/inappropriate landings' by pest insects of cruciferous plants. Entomologia Experimentalis et Applicata 96: 91-102.

Frearson DJT, Ferguson AW, Campbell JM & Williams IH (2005) The spatial dynamics of pollen beetles in relation to inflorescence growth stage of oilseed rape: implications for trap crop strategies. Entomologia Experimentalis et Applicata 116: 21-29.

Free JB & Williams IH (1978) The response of the pollen beetle, *Meligethes aeneus*, and the seed weevil, *Ceutorhynchus assimilis*, to oil-seed rape, *Brassica napus*, and other plants. Journal of Applied Ecology 15: 761-774.

Giamoustaris A & Mithen R (1995) The effect of modifying glucosinolate content of leaves of oilseed rape (*Brassica napus* ssp. *oleifera*) on its interaction with specialist and generalist pests. Annals of Applied Biology 126: 347-363.

Giamoustaris A & Mithen R (1996) The effect of flower colour and glucosinolates on the interaction between oilseed rape and pollen beetles. Entomologia Experimentalis et Applicata 80: 206-208.

Godan D (1951) Über Nahrungs- und Brutpflanzen des Rapserdflohs (*Psylliodes chrysocephala* L.). Anzeiger für Schädlingskunde 24: 81-84.

Hannunnen S (2005) Modelling the interplay between pest movement and the physical design of trap crop systems. Agricultural and Forest Entomology 7: 11-20.

Kalischuk AR & Dosdall LM (2004) Susceptibilities of seven Brassicaceae species to infestation by the cabbage seedpod weevil (Coleoptera : Curculionidae). Canadian Entomologist 136: 265-276.

Kaufmann O (1941) Zur Biologie des Rapserdflohes (*Psylliodes Chrysocephala* L.). Zeitschrift für Pflanzenkrankheiten (Pflanzenpathologie) und Pflanzenschutz 51: 305-324.

Koritsas VM, Lewis JA & Fenwick GR (1989) Accumulation of indole glucosinolates in *Psylliodes chrysocephala* L. infested, or *Psylliodes chrysocephala* L. damaged tissues of oilseed rape (*Brassica napus* L.). Experientia 45: 493-495.

Koritsas VM, Lewis JA & Fenwick GR (1991) Glucosinolate response of oilseed rape, mustard and kale to mechanical wounding and infestation by cabbage stem beetle (*Psylliodes chrysocephala*). Annals of Applied Biology 118: 209-221.

Lamb RJ 1980. Hairs protect pods of mustard (Brassica hirta Gisilba) from flea beetle feeding damage. Canadian Journal of Plant Science 60, 1439-1440.

Lambdon PW, Hassall M & Mithen R (1998) Feeding preferences of woodpigeons and flea-beetles for oilseed rape and turnip rape. Annals of Applied Biology 133: 313-328.

Linz B (1991) Hymenopteren als Parasitoide des Rapsglanzkäfers (*Meligethes* spp., Nitidulidae) und des Kohlschotenrüsslers (*Ceutorhynchus assimilis*, Curculionidae) auf Raps und Rübsen. Mitteilungen der Deutschen Gesellschaft für Allgemeine und Angewandte Entomologie 8: 90-92.

Meisner J & Mitchell BK (1984) Phagodeterrency induced by some secondary plant substances in adults of the flea beetle *Phyllotreta striolata*. Journal of Plant Diseases and Protection 91: 301-304.

Mewis I, Rohr F, Tokuhisa JG & Ulrichs C (2008) Auswirkung einer Modifizierung des Glucosinolatprofils von *Arabidopsis thaliana* (L.) auf verschieden spezialisierte Lepidoptera. Mitteilungen der Deutschen Gesellschaft für Allgemeine und Angewandte Entomologie 16: 167-169.

Mithen R (2001) Glucosinolates - biochemistry, genetics and biological activity. Plant Growth Regulation 34: 91-103.

Nilsson C (2004) Trap plants to avoid insecticide application against pollen beetles in oilseed rape. IOBC/wprs Bulletin 27 (10): 215-221.

Pöykkö H (2006) Females and larvae of a geometrid moth, *Cleorodes lichenaria*, prefer a lichen host that assures shortest larval period. Environmental Entomology 35: 1669-1676.

Rohr F, Ulrichs C & Mewis I (2009) Variability of aliphatic glucosinolates in *Arabidopsis thaliana* (L.) - Impact on glucosinolate profile and insect resistance. Journal of Applied Botany and Food Quality 82: 131-135.

Root (1973) Root, R. B., 1973. Organization of a plant-arthropod association in simple and diverse habitats: the fauna of collards (*Brassica oleracea*). Ecological Monograph 43: 95–124. Cited in: Finch S & Collier RH (2000) Host-plant selection by insects - a theory based on 'appropriate/inappropriate landings' by pest insects of cruciferous plants. Entomologia Experimentalis et Applicata 96: 91-102

Schoonhoven LM, van Loon JJA & Dicke M (2005) Insect-plant biology. Oxford University Press, Oxford: 421pp

Schulz R-R (1985) Untersuchungen zur Vermehrungsrate des Rapserdflohs (*Psylliodes chrysocephala* L.). Archiv für Pflanzenpathologie und Pflanzenschutz 21: 305-311.

Shelton AM & Badenes-Perez E (2006) Concepts and applications of trap cropping in pest management. Annual Review of Entomology 51: 285-308.

Shikano I, Akhtar Y & Isman MB (2010) Relationship between adult and larval host plant selection and larval performance in the generalist moth, *Trichoplusia ni.* Arthropod-Plant Interactions 4: 197-205.

Siemens DH & MitchellOlds T (1996) Glucosinolates and herbivory by specialists (Coleoptera: Chrysomelidae, Lepidoptera: Plutellidae): Consequences of concentration and induced resistance. Environmental Entomology 25: 1344-1353.

Strauch S (2010) Regulierung des Rapsglanzkäfers im ökologischen Landbau. VDM Verlag DR. Müller Aktiengesellschaft & Co. KG, Saarbrücken. 82pp

Tansey JA, Dosdall LM, Keddie A, Fletcher RS & Kott LS (2010) Antixenosis and antibiosis resistance to *Ceutorhynchus obstrictus* in novel germplasm derived from *Sinapis alba* x *Brassica napus*. Canadian Entomologist 142: 212-221.

Tansey JA & Dosdall LM (2011) Differential responses by some insect pests to novel insect-resistant *Brassica napus* L.: 13th International Rapeseed Congress, Prague.

Theunissen J, Booij CJH & Lotz LAP (1995) Effects of intercropping white cabbage with clovers on pest infestation and yield. Entomologia Experimentalis et Applicata 74: 7-16.

Thioulouse J (1987) Space-time structures in a winter rape pest population, *Psylliodes chrysocephala* (Col. Chrysomelidae) methodological proposals and biological interpretations. Journal of Applied Ecology 24: 435-450.

Tukahirwa EM & Coaker TH (1982) Effect of mixed cropping on some insect pests of brassicas; reduced *Brevicoryne brassicae* infestations and influences on epigeal predators and the disturbance of oviposition behaviour in *Delia brassicae*. Entomologia Experimentalis et Applicata 32: 129-140.

Uddin MM, Ulrichs C & Mewis I (2008) *Phaedon cochleariae* (F.) performance on different crucifer varieties with different glucosinolate profiles. Communications in Agricultural and Applied Biological Sciences 73: 563-572.

Uddin MM, Ulrichs C, Tokuhisa JG & Mewis I (2009) Impact of glucosinolate structure on the performance of the crucifer pest *Phaedon cochleariae* (F.). Journal of Applied Botany and Food Quality 82: 108-113.

Ulber B, Williams IH, Klukowski U, Luik A & Nilsson C (2010) Parasitoids of oilseed rape pests in Europe: key species for conservation biocontrol. In: Biocontrol-based integrated management of oilseed rape pests (ed. IH Williams) Springer Science+Business Media B.V., Dordrecht, Heidelberg, London, New York, pp. 45-76.

Ulmer BJ & Dosdall LM (2006) Glucosinolate profile and oviposition behavior in relation to the susceptibilities of Brassicaceae to the cabbage seedpod weevil. Entomologia Experimentalis et Applicata 121: 203-213.

Uppström B (1995) Seed chemistry. In: Brassica oilseeds - production and utilization (ed. DS Kimber & DI McGregor). CAB International, Wallingford, pp. 217-242.

Williams IH (1989) Pest incidence on single low and double low oilseed rape. Aspects of Applied Biology 23: 277-286.

Williams IH (2004) Advances in insect pest management of oilseed rape in Europe. In: Insect pest management (ed. AR Horowitz & I Ishaaya). Springer-Verlag, Berlin Heidelberg, pp. 181-208.

Williams IH & Bartlet E (1993) Host plant affecting the feeding of cabbage stem flea beetle (*Psylliodes chrysocephala* L.) on oilseed rape. IOBC/wprs Bulletin 16 (9): 202-209.

Williams IH & Free JB (1978) The feeding and mating behaviour of pollen beetles (*Meligethes aeneus* Fab.) and seed weevils (*Ceutorhynchus assimilis* Payk.) on oil-seed rape (*Brassica napus* L.). Journal of Agricultural Science 91: 453-459.

Williams IH, Ferguson AW, Kruus M, Veromann E & Warner DJ (2010) Ground beetles as predators of oilseed rape pests: incidence, spatio-temporal distributions and feeding. In: Biocontrol-based integrated management of oilseed rape pests (ed. IH Williams). Springer Science+Business Media B.V., Dordrecht, Heidelberg, London, New York: 115-150

Abstract

Substantial economic yield losses are regularly caused by insect pests to crops of oilseed rape (*Brassica napus* L.) in Europe. The control of these pests is usually achieved by application of synthetic insecticides. Emerging resistances of insects against specific active ingredients require integrated pest management strategies. The aim of this study was to identify cultivars of oilseed rape and other brassicaceous species which are less susceptibility or even resistant to herbivore feeding. Moreover, the suitability of turnip rape as a trap crop for insect pests in oilseed rape was investigated. Experiments were conducted under controlled climate conditions to assess the larval performance and adult feeding preference of the cabbage stem flea beetle (*Psylliodes chrysocephala* L.; CSFB) on different brassicaceous species. To identify potential resistance mechanisms, leaf laminas were analyzed for sugars, phenolics and glucosinolates and petioles were analyzed for glucosinolates. Additionally, trichome densities and dry weight of leaf laminas was assessed. Extensive field plot experiments were established to evaluate the suitability of different turnip rape (*Brassica rapa* L. *silvestris*) cultivars as trap crop. The effect of turnip rape, grown as perimeter strips or in mixtures with oilseed rape, on infestation by CSFB, rape stem weevil (*Ceutorhynchus napi* Gyll.), cabbage stem weevil (*Ceutorhynchus pallidactylus* Marsh.), pollen beetle (*Meligethes aeneus* Fab.) and cabbage seed weevil (*Ceutorhynchus obstrictus* Marsh.) on oilseed rape was investigated.

Larval performance of CSFB significantly differed between brassicaceous species. While larvae reared in turnip rape gained the highest weight, larvae developing on white mustard (*Sinapis alba* L.) gained lowest. Larval performance was also significantly reduced when larvae fed on white cabbage (*Brassica oleracea* L. var. *capitata* f. *alba*) and oilseed radish (*Raphanus sativa* L. var. *oleiformis*) compared to oilseed rape. Performance of larvae was positively correlated with the content of the glucosinolates progoitrin and 4-hydroxybrassicin in the leave petioles. In dual choice experiments adult CSFB preferred turnip rape to oilseed rape for feeding, while white cabbage was less preferred. Feeding preference was not correlated with the content of glucosinolates, sugars and penolics in the leaf lamina while the feeding preference was positively correlated with the trichome density. In field experiments turnip rape was significantly higher infested by CSFB and pollen beetle than oilseed rape. Among four cultivars of turnip rape 'Malwira' was preferred by pollen beetles. However, higher attractiveness of the turnip rape in perimeter strips and in mixtures with oilseed rape did not reduce the infestation levels of the main crop by pests.

The antixenosis resistance of white mustard to CSFB larvae is regarded as having potential for breeding oilseed rape cultivars less susceptible to insect pests. Both trap cropping strategies did not reduce pest infestation of the main crop and are not suitable for use as a single pest management strategy. A combination of a turnip rape perimeter strips and targeted insecticide application to these strips may contribute to an effective trap crop strategy. Furthermore a less susceptible cultivar of oilseed rape as main crop in combination with a highly attractive trap crop could contribute to a better efficacy of the perimeter strips for insect pest management.

Zusammenfassung

Raps ist die wichtigste Ölpflanze in Deutschland, jedoch ist der Ertrag stark durch tierische Schädlinge gefährdet. Die Kontrolle dieser Schädlinge erfolgt meist mittels synthetischer Insektizide. Da jedoch eine Häufung von Schädlingspopulationen mit Resistenzen gegen wichtige Insektizidgruppen zu beobachten ist, wird die Suche nach integrierten Bekämpfungsstrategien zunehmend wichtiger. Ziel dieser Studie war es, Brassicaceen mit verminderter Anfälligkeit gegen die wichtigsten Schadinsekten zu finden. Des weiteren wurde die Eignung von Winterrübsen (*Brassica rapa* L. *silvestris*) als Fangpflanze in Winterraps untersucht. In Klimakammerversuchen wurde die Larvenentwicklung und das Fraßverhalten von adulten Rapserdflöhen (*Psylliodes chrysocephala* L.) an unterschiedlichen Brassicaceen untersucht. Um Resistenzmechanismen in den verschiedenen Pflanzenarten zu identifizieren, wurden Glucosinolate, Zucker und Gesamtphenole in den Blattspreiten analysiert, sowie die Trichomdichte auf der Blattoberseite und das Trockengewicht bestimmt. In den Blattstielen, in denen sich die Rapserdflohlarven natürlicherweise entwickeln, wurde ebenfalls der Glucosinolatgehalt bestimmt. In dreijährigen Feldversuchen wurde die Effizienz verschiedener Rübsensorten als mögliche Fangpflanze an zwei Standorten untersucht. Außerdem wurde die Eignung von Rübsen-Fangstreifen sowie einer Mischsaat aus Raps und Rübsen als Maßnahme zur Minderung des Rapserdfloh- (*Psylliodes chrysocephala* L.), Rapsstengelrüssler- (*Ceutorhynchus napi* Gyll.), Kohltriebrüssler- (*Ceutorhynchus pallidactylus* Marsh.), Rapsglanzkäfer- (*Meligethes aeneus* Fab.) und Kohlschotenrüsslerbefalls (*Ceutorhynchus obstrictus* Marsh.) in Winterraps untersucht.

Die Entwicklung der Rapserdflohlarven unterschied sich signifikant in den untersuchten Wirtspflanzenarten. Während Larven, die an Rübsen angezogen wurden, das höchste Frischgewicht aufwiesen, waren Larven, die an Weißem Senf (*Sinapis alba* L.) angezogen wurden, am leichtesten. Das Larvengewicht war an Weißkohl (*Brassica oleracea* L. var. *capitata* f. *alba*) und Ölrettich (*Raphanus sativa* L. var. *oleiformis*) im Vergleich zu Raps ebenfalls signifikant reduziert. In unseren Versuchen konnten wir eine positive Korrelation zwischen dem Gewicht der Rapserdflohlarven und den Glucosinolaten Progoitrin und 4-Hydroxybrassicin in den Blattstielen der Wirtspflanzen feststellen. In den Wahlversuchen bevorzugten die adulten Rapserdflöhe Rübsen signifikant vor Raps. Weißkohl wies hingegen einen geringeren Blattfraß als Raps auf, wenn beide Arten zur Wahl angeboten wurden. Die konsumierte Blattfläche der Versuchstiere war positiv mit der Trichomdichte korreliert, jedoch nicht mit dem Glucosinolat-, Zucker- und Gesamtphenolgehalt oder dem Trockengewicht der Blattspreiten. In den Feldversuchen waren die Rübsensorten signifikant stärker mit Rapserdflohlarven und Rapsglanzkäfern befallen als der Winterraps, wobei die Rübsensorte ‚Malwira' attraktiver für den Rapsglanzkäfer war als die übrigen geprüften Rübsensorten. Die höhere Attraktivität des Rübsens führte jedoch weder in den Fangstreifen- noch in den Mischsaatversuchen zu einer Reduktion des Schädlingsbefalls im angrenzenden Rapsbestand beziehungsweise in den Rapspflanzen der Mischsaat.

Die geringe Wirtsqualität des Weißen Senf für die Larven des Rapserdflohs könnte zukünftig für die Züchtung von Rapssorten mit geringerer Anfälligkeit gegenüber Insekten genutzt werden. Weitere Untersuchungen müssen zeigen, ob eine Kombination aus Rübsen-Fangstreifen und einer Randbehandlung mit Insektiziden die Effektivität des Fangpflanzenverfahrens erhöhen kann.

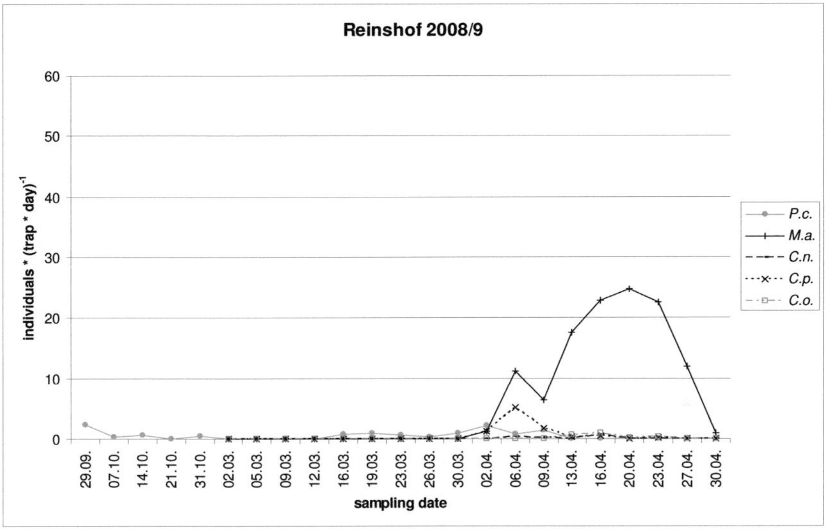

Fig. 1: Numbers of *Psylliodes chrysocephala*, *Meligethes aeneus*, *Ceutorhynchus napi*, *Ceutorhynchus pallidactylus* and *Ceutorhynchus obstrictus* caught in yellow water traps at Göttingen 2008/09 (16.09.-30.10.2008 and 27.02.-30.04.2009); mean of four traps; mean number of individuals per day caught between two sampling dates

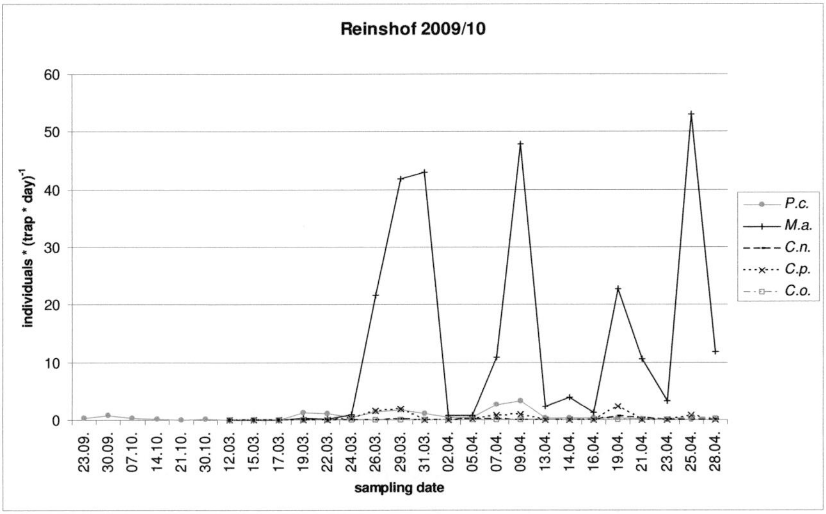

Fig. 2: Numbers of *Psylliodes chrysocephala*, *Meligethes aeneus*, *Ceutorhynchus napi*, *Ceutorhynchus pallidactylus* and *Ceutorhynchus obstrictus* caught in yellow water traps at Göttingen 2009/10 (16.09.-30.10.2009 and 10.03.-28.08.2010); mean of four traps; mean number of individuals per day caught between two sampling dates

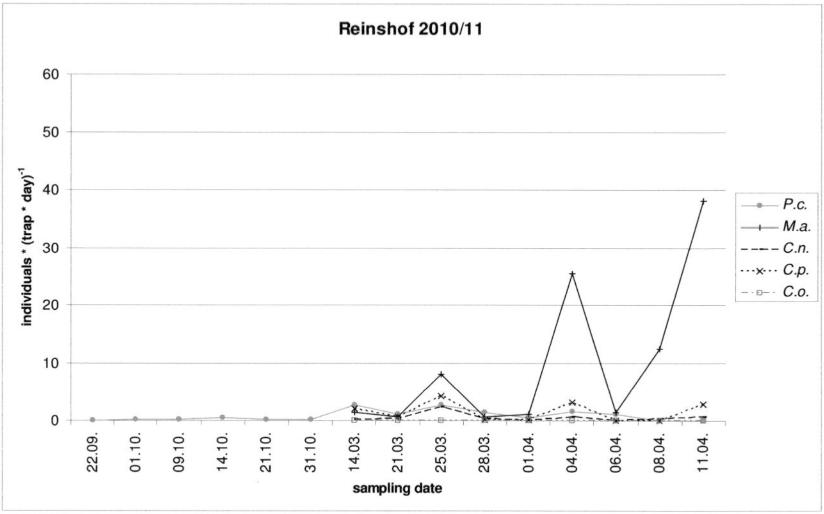

Fig. 3: Numbers of *Psylliodes chrysocephala*, *Meligethes aeneus*, *Ceutorhynchus napi*, *Ceutorhynchus pallidactylus* and *Ceutorhynchus obstrictus* caught in yellow water traps at Göttingen 2010/11 (14.09.-31.10.2010 and 10.03.-11.04.2011); mean of four traps; mean number of individuals per day caught between two sampling dates

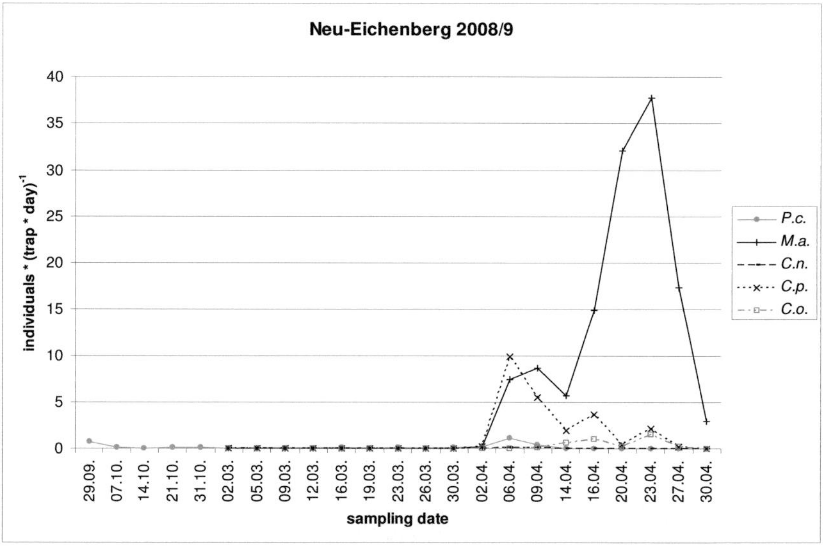

Fig. 4: Numbers of *Psylliodes chrysocephala*, *Meligethes aeneus*, *Ceutorhynchus napi*, *Ceutorhynchus pallidactylus* and *Ceutorhynchus obstrictus* caught in yellow water traps at Neu-Eichenberg 2008/09 (16.09.-31.10.2008 and 27.02.-30.04.2009); mean of four traps; mean number of individuals per day caught between two sampling dates

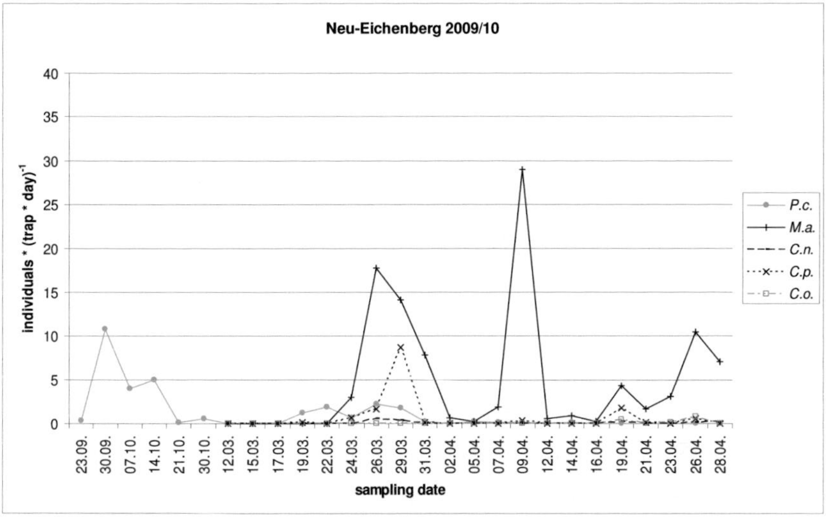

Fig. 5: NUmbers of *Psylliodes chrysocephala*, *Meligethes aeneus*, *Coutorhynchus napi*, *Ceutorhynchus pallidactylus* and *Ceutorhynchus obstrictus* caught in yellow water traps at Neu-Eichenberg 2009/10 (16.09.-30.10.2009 and 10.03.-28.04.2010); mean of four traps; mean number of individuals per day caught between two sampling dates

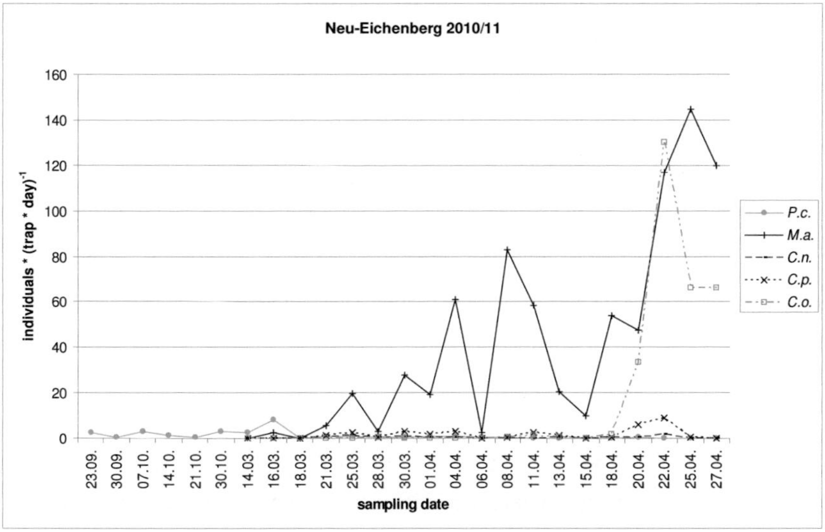

Fig. 6: Numbers of *Psylliodes chrysocephala*, *Meligethes aeneus*, *Ceutorhynchus napi*, *Ceutorhynchus pallidactylus* and *Ceutorhynchus obstrictus* caught in yellow water traps at Neu-Eichenberg 2010/11 (15.9.-30.10.2010 and 10.03.-27.04.2011); mean of four traps; mean number of individuals per day caught between two sampling dates

Danksagung

Prof. Dr. Stefan Vidal danke ich für die Aufnahme in seine Arbeitsgruppe und dass er sich bereit erklärte, als Gutachter für meine Arbeit zu fungieren.

Prof. Dr. Heiko Becker möchte ich sehr herzlich für seine Bereitschaft danken, als Korreferent zu fungieren.

Ein besonderer Dank gilt Herrn Dr. Bernd Ulber für seine schier unendliche Geduld, meine Fragen und Probleme mit mir zu diskutieren und für seine sehr konstruktive Kritik.

Rainer Wedemeyer danke ich für seine Hilfe bei den Feldversuchen; ohne seine Erfahrung wären viele Versuche misslungen.

Dr. Helmut Saucke danke ich für die Möglichkeit, die auf dem Versuchsgut in Neu-Eichenberg erhobenen Daten für meine Arbeit zu verwenden.

Dr. Inga Smit und Gunda Jansen danke ich für die Möglichkeit, Zucker und Gesamtphenole in meinen Proben zu bestimmen.

Herrn Uwe Ammermann und Dr. Christian Möllers danke ich für die Unterstützung bei der Analyse der Glucosinolate.

Ein besonderer Dank gilt Ines Vollhardt für ihre Bereitschaft, Teile meiner Arbeit Korrektur zu lesen.

Des weiteren bedanke ich mich bei den vielen Helfern, die mir während der letzten drei Jahre beim Unkrauthacken, in statistischen Fragen oder bei der Pflege meiner Gewächshauspflanzen geholfen haben: Dorothea Mennerich, Katja Fischer, Marie-Luise Tölle, Helge Stahlmann, Antje Reinhardt, Anne Wilstermann, Michael Eickermann, Henrike Hennies, Verena Werner, Michael Klingenberg, Kathrin Agethen, Eugen Hodyl, sowie einer Vielzahl studentischer Hilfskräfte, Praktikanten und Zivildienstleistender.

Meinen Eltern, meiner Schwester und meiner Freundin möchte ich dafür danken, dass sie mich jederzeit unterstützt haben und auch in schwierigen Zeiten zu mir gehalten haben.

Curriculum vitae

Alexander Döring

date of birth	November 17th, 1982
place of birth	Fritzlar (Germany)

08/1993 - 06/2002	grammar school (König-Heinrich-Schule, Fritzlar) with "Allgemeine Hochschulreife" as final exam
08/2002 - 05/2003	civilian service at „Baunataler Werkstätten"
04/2004 - 03/2007	studies in agricultural sciences at the Georg-August-University Goettingen, final exam: Bachelor of Science
04/2007-12/2008	studies in agricultural sciences at the Georg-August-University Goettingen, final exam: Master of Science
since 01/2009	Ph. D. student at the Georg-August-University Goettingen, Department of Crop Sciences, Division of Agricultural Entomology